GCSE MATHEMATICS

LONGMAN
REVISE
GUIDES

Longman GCSE Guides

SERIES EDITORS:
Geoff Black and Stuart Wall

TITLES AVAILABLE:
Biology
Business Studies
Chemistry
English
English Literature
Mathematics
Physics
World History

FORTHCOMING:
Art and Design
British and European History
Computer Studies
Commerce
C.D.T.
  Design and Realisation
  Design and Technology
Economics
French
Geography
German
Home Economics
Religious Studies

LONGMAN
REVISE
GUIDES

GCSE

MATHEMATICS

Brian Speed

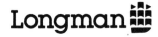

Longman Group UK Limited,
Longman House, Burnt Mill, Harlow,
Essex CM20 2JE, England
and Associated Companies throughout the world.

First published 1988

British Library Cataloguing in Publication Data

Speed, Brian
    Mathematics. — (Longman GCSE revise guides).
    1. Mathematics — Study and teaching
    (Secondary) — Great Britain 2. General Certificate
    of Secondary Education — Study guides
    I. Title
    510'.76 QA14.G7

ISBN 0-582-02431-5

Illustrated by Sharon Fry

Set in 9/12pt Century Book Roman

Printed and bound in Great Britain by
Thamesmouth Printing Group, Basildon, Essex.

C O N T E N T S

EDITORS' PREFACE

Longman Revise Guides are written by experienced examiners and teachers, and aim to give you the best possible foundation for success in examinations and other modes of assessment. Much has been said in recent years about declining standards and disappointing examination results. While this may be somewhat exaggerated, examiners are well aware that the performance of many candidates falls well short of their potential. The books encourage thorough study and a full understanding of the concepts involved and should be seen as course companions and study guides to be used throughout the year. Examiners are in no doubt that a structured approach in preparing for examinations and in presenting course-work can, together with hard work and diligent application, substantially improve performance.

The largely self-contained nature of each chapter gives the book a useful degree of flexibility. After starting with Chapters 1 and 2, all other chapters can be read selectively, in any order appropriate to the stage you have reached in your course. We believe that this book, and the series as a whole, will help you establish a solid platform of basic knowledge and examination technique on which to build.

A C K N O W L E D G E M E N T S

I would like to thank the following people for their valuable contribution to the production of this book: Stuart Wall and Geoff Black for their editing and most useful suggestions at each stage of production; my fifth form pupils at Pope Pius X school (87/88) for their help with the questions; my mother Mrs Elsie Speed for her devoted typing and spelling corrections of the original manuscript; Gillian my wife, who very patiently looked after our children while I worked; and finally to my lads James, John and Joseph for their patient understanding of Dad 'not being able to play again tonight!'

I am also indebted to the following Examination Boards for giving me permission to use some of their specimen GCSE questions in this book:

Northern Examination Association	(NEA)
Midlands Examination Group	(MEG)
London and East Anglian Group	(LEAG)
Southern Examination Group	(SEG)

The above groups do not accept responsibility for the answers I have given to their specimen questions, and so all suggestions and any mistakes in these answers are entirely my responsibility. I will be most grateful to any reader who informs me of any such mistakes should they occur.

Brian Speed 1988

GCSE
IN
MATHEMATICS

From 1988 the GCSE examination in mathematics replaces the 'O' level, the CSE and the 16+ examinations. It has come about after many years of discussion and development within education as a whole, and not just within mathematics. In this chapter we consider the grading system in GCSE mathematics, the different levels of examination, the importance of coursework and the detailed requirements of each Examination Board.

LEAG
NEA
MEG
SEG
WJEC
NISEC
IGCSE

G R A D E S

The 'O' level grades and the CSE grades were used as levels of *attainment* rather than as a guide to the *knowledge* that was needed to obtain any given grade. Now, with GCSE, the examinations are linked to 'Grade Criteria'. These are descriptions of what each grade means in terms of *knowledge* and *skills*. The GCSE in mathematics is *not* a system of hurdles. You will be able to obtain a particular grade even if you cannot do everything in that grade's description, since a weakness in one aspect can be *balanced* by an above-average performance in another.

The *grade descriptions* (criteria) have been written by the SEC (Secondary Examination Council), which has overall responsibility for maintaining standards across the different Examination Boards. Much of the generally accepted *informal* ideas that lay behind the 'O' level and CSE grades have become part of the *formal* grade descriptions. For each syllabus in mathematics there are just three grade descriptions at present, namely those for grades A, C and F. Although, at the time of writing, *full* descriptions of the other grades have not yet been published by the SEC, it has been possible to devise grade *checklists for each topic*. These are presented at the end of each chapter and give you a very good guide as to what you need to *know* and be able to *do* for grades A, C and F in that topic. You can therefore build up a picture of what each of these grades now means. As a guide, the table indicates the relationship between the GCSE grades and the previous 'O' level and CSE grades. Note that 'U' means 'unclassified', or rather, not meeting the requirements for the entry level.

GCSE	A	B	C	D	E	F	G	U
'O' level	A	B	C	D	D/E	E/U	U	U
CSE	–	–	1	2	3	4	5	U

L E V E L   O F   E N T R Y

It is an important feature of the GCSE that pupils must not be required to prepare for examinations which are unsuited to their level of attainment; nor must these examinations be of a kind that will undermine the confidence of pupils. Hence in mathematics you will find three levels of entry: a *basic*, an *intermediate* and a *higher* level. These levels have different labels in each Examination Board, but they all mean the same thing.

Examination papers are written at *each level* and each level is targeted at three grades:

Level	Target
Basic	E, F, G
Intermediate	C, D, E (F usually available)
Higher	A, B, C (D usually available)

If you are entered at the *intermediate level*, the highest grade you can obtain is a C, and the lowest an F. If you fall below the standard for a grade F you will be unclassified. Similarly at the *higher level*, if you fall below the standard for a grade D you will be unclassified. It is vital then that you are entered at the *correct* level, since you cannot be entered for two levels.

To satisfy the 'confidence of pupils' aspect, the papers are written with a particular end in view. Namely to provide enough scope for the more able students *at that level* to do well and to obtain the higher target grades, *but* to set papers which will allow the *lowest target grade* to obtain at least 50% of the marks. So, if you are correctly entered, you can expect to be able to do *at least half* the examination paper.

C O U R S E W O R K

Another innovation over the years, particularly in CSE mathematics, has been to allow coursework to contribute to the final grade. This is now going to be *compulsory* for *all* mathematics syllabuses after 1991, but until 1991 it will be optional. Some Examination Boards have decided that their syllabuses will *not* use coursework before 1991, while others have insisted on coursework from the start, and yet others allow coursework to be selected as an *option*. You should of course have been told by your teacher at school or college as to whether coursework is to be included in *your* final grade. In the following section you should be able to locate your Examination Board and syllabus to check the type of assessment you will face.

M O D E S

There are three different modes of GCSE examinations

MODE 1   This is the mode where all the written papers are set by the *Examination Board itself* based upon the syllabuses they have devised. The vast majority of the examinations that are mentioned in this book are Mode 1.

MODE 2   This is the mode where the Examination Board writes the examination papers, *but* they are based on a syllabus devised by an individual school or college (or group of schools/colleges). If your centre is operating a Mode 2 examination, then you need to ask for a syllabus so that you can make your own list of relevant topics.

MODE 3   This is the mode where a centre or group of centres write their *own examination papers*, which are based upon syllabuses which *they have themselves devised*. If your centre is operating a Mode 3 examination, then you need to ask for the syllabus so that you can make your own list of relevant topics. Often a Mode 3 examination will ask for more course work than the other modes. Again, it is up to you to find out so that you are fully prepared.

S C H E M E S   O F   A S S E S S M E N T

1 ⟩ LONDON & EAST ANGLIAN GROUP (LEAG)

Syllabus A (no coursework up to 1990)

Level X (basic)	will sit Paper 1 and Paper 2
Level Y (intermediate)	will sit Paper 2 and Paper 3
Level Z (higher)	will sit Paper 3 and Paper 4

There is no choice of questions on each paper, and each paper is worth 50% of the final total.

Syllabus B

The papers are set as above, but coursework is now included. Each paper is worth 35% and the coursework is worth 30%. The coursework will consist of a mental test worth 5% and five set tasks, each of 5%, give the other 25%.

Syllabus SMP (no coursework up to 1990)

The papers are set as for Syllabus A, with no coursework and with each paper worth 50%.

2 ⟩ NORTHERN EXAMINING ASSOCIATION (NEA)

Syllabuses A and B

Both include an optional coursework component (until 1991 when it will be compulsory).

Scheme I (with no coursework)

Level P (basic)	will sit Paper 1 and Paper 2 (both 50%)
Level Q (intermediate)	will sit Paper 2 (45%) and Paper 3 (55%)
Level R (higher)	will sit Paper 3 (45%) and Paper 4 (55%)

There is no choice of questions on any of the papers

Scheme II

There is coursework worth 25% of the assessment. The combination of written papers is the same as above, but:

Level P	–both $37\frac{1}{2}$%
Level Q	–Paper 2, 33%; Paper 3, 42%
Level R	–Paper 3, 33%; Paper 4, 42%

Syllabus C

This has compulsory coursework worth 25%, with the combination of papers the same as for Syllabuses A and B, but the weightings are:

Level P	–both $37\frac{1}{2}$%
Level Q	–Paper 2, 34%; Paper 3, 41%
Level R	–Paper 3, 34%; Paper 4, 41%

3 MIDLAND EXAMINING GROUP (MEG)

Syllabus Mathematics

Has an optional coursework component until 1991, after which coursework will be compulsory.

Scheme I (with no coursework)

Each level will take two papers, both worth 50%. The first paper will be short-answer questions, all answered on the question paper, and no choice. The second paper contains longer structured questions and offers choice in section B at the intermediate and the higher levels. The intermediate choice is 4 out of 5 questions. The higher choice is 4 out of 6 questions.

At the foundation (lower) level, the answers will be written on the question paper, and at the intermediate and higher levels they will be written on separate script paper.

Scheme II (with coursework worth 25%)

The papers are the same as for Scheme I, but the weightings are now: first paper 50%; second paper 25%. The coursework assignment will be the candidate's choice of five assignments arranged with the school or college.

Syllabus Mathematics (mature)

The type of papers and choice is the same as for the syllabus 'mathematics'. Candidates take two papers suitable for their level, with the weightings being: first paper, 40%; second paper, 50%. However, for the 'mature' syllabus there is an aural test giving the other 10%. The aural test will be 20 questions, to be answered on an answer sheet provided. You are **NOT** allowed to use a calculator in the aural test.

Syllabus Mathematics (SMP (11–16))

There will be three parts to the assessment scheme at each level:

(a) Written papers

Foundation level	Paper 1 and Paper 2, 35% each
Intermediate level	Paper 2 and Paper 3, 35% each
Higher level	Paper 3 and Paper 4, 35% each

(b) Coursework, giving 25%, consisting of eight tasks provided by the Examination Board.

(c) Aural test, giving 5%. This will consist of two tests given orally from questions provided by the Examination Board.

Syllabus Mathematics (SMP) (until 1991)

There is no coursework component, and the combination of papers is exactly the same as for the syllabus SMP(11–16), except that the weightings will be 50% for each paper. Candidates will be required to attempt all questions, except in Paper 4, section B, where there will be a choice of 4 out of 6 questions.

4 ▷ SOUTHERN EXAMINING GROUP (SEG)

There are two schemes, one with coursework and one (until 1991) with no coursework.

Scheme I (with no coursework)
This consists of two parts:

(a) Level 1 (basic)   will sit Paper 1 and Paper 2 (45% each)

Level 2 (intermediate)   will sit Paper 2 and Paper 3 (45% each)

Level 3 (higher)   will sit Paper 3 and Paper 4 (45% each)

There is no choice of questions on any of the papers.

(b) Aural tests giving 10%. Each candidate will take two 15-minute aural tests as:

Level 1 (basic)   Aural tests 1 and 2

Level 2 (intermediate)   Aural tests 2 and 3

Level 3 (higher)   Aural tests 3 and 4

Calculators are **NOT** allowed to be used for the Aural tests.

Scheme II (with coursework)
This consists of three parts:

(a) The written papers are the same combinations as above, but the weightings are now 25% each.

(b) The Aural tests as above give 10%.

(c) The coursework, or 'centre-based assessment' as it is called in the syllabus, gives 40%. This will consist of three units, which will be arranged by you and your school or college.

5 ▷ WELSH JOINT EDUCATION COMMITTEE (WJEC)

There are two schemes, one with coursework and one without.

Scheme A (with coursework)
This will consist of two parts, written papers and coursework, with weightings in the ratio of 200 : 70.

(a) The written papers each earn 100 marks and are taken as:

Level 1 (basic)   will sit Paper 1 and Paper 2

Level 2 (intermediate)   will sit Paper 2 and Paper 3

Level 3 (higher)   will sit Paper 3 and Paper 4

(b)   The coursework will consist of two tasks provided by the Examination Board, and combined to give 70 marks.

Scheme B (without coursework)
The written papers will be taken as in Scheme A and each one is worth 50%. There is no choice of questions on any paper.

6 ▷ NORTHERN IRELAND SCHOOLS EXAMINATIONS COUNCIL (NISEC)

There are two syllabuses A and B. The syllabus content and the written papers are identical in each case, the only difference being that syllabus B has no coursework included in the assessment.

Syllabus A

There are *three* parts to the assessment of Syllabus A at each level.

(a) Written papers
At each level you sit *two written papers*, each of which will consist of *short answer* questions and *long questions* (most of which will be *structured*). There is no choice of question offered. Each paper is worth 35% of the final assessment.

(b) Aural and computation
There will also be an Aural and computation test, which is set for each level. This will test your mental arithmetic and how well you can understand a spoken instruction regarding information available on a separate document. This test is worth 10% of the assessment.

(c) Coursework
You will normally have to hand in *four* assignments for assessing. Your teacher will tell you what these assignments are. They could include work on topics such as practical geometry, measurement, statistics, everyday application of mathematics and investigations. This coursework element is worth 20% of the assessment.

Syllabus B

The same as syllabus A *except* that there is no coursework to include. The combinations are therefore different and as follows:

(a) the two written papers are each worth 45% of the assessment.

(b) the aural and computation test is worth 10% of the assessment.

---

7 ▷ INTERNATIONAL GENERAL CERTIFICATE OF SECONDARY EDUCATION (IGCSE)

This syllabus has been designed to meet international mathematical needs while being based on the United Kingdom's national criteria as published by the SEC.

There are only *two* levels available: the basic level being included in the *lower level* of the two which is called the *core curriculum* where the only grades available are C to G; the *higher level* being called the *extended curriculum*, where the only available grades are from A to E. There is also an *optional* coursework element in place of part of the written papers. The assessment will be in *three* parts:

(a) a *written* paper of *short answer* questions.
(b) a *written* paper of *structured questions*.
(c) a *written* paper of *problems* **or** the *school based assessment*.

There is *no* choice of question on any paper. The combination of the different parts of the assessment are:

core:	(a) first paper	35%
	(b) second paper	40%
	(c) third part	25%
extended:	(a) first paper	37.5%
	(b) second paper	37.5%
	(c) third part	25%

The *school based assessment* which is *optional* consists of four coursework assignments (20%) and two aural tests (5%). The four coursework assignments will be on the four areas of:

- statistics and/or probability
- geometry
- investigations
- practical applications of mathematics.

The aural tests will be about fifteen minutes of single response questions aimed at each different level.

You may use a suitable calculator in each part of the assessment and, if you wish, four figure tables also. For centres that are in areas where electronic calculators are not readily and cheaply obtainable, there is an alternative version of the examination available.

EXAMINATION BOARDS: ADDRESSES

1 ▷ LONDON AND EAST ANGLIAN GROUP (LEAG)	London	University of London Schools Examinations Board
		Stewart House, 32 Russell Square, London WC1B 5DN
	LREB	London Regional Examinations Board
		Lyon House, 104 Wandsworth High Street, London SW18 4LF
	EAEB	East Anglian Examinations Board
		The Lindens, Lexden Road, Colchester, Essex CO3 3RL
2 ▷ NORTHERN EXAMINATION ASSOCIATION (NEA)	JMB	Joint matriculation Board
		Devas Street, Manchester M15 6EU
	ALSEB	Associated Lancashire Schools Examining Board
		12 Harter Street, Manchester M1 6HL
	NREB	North Regional Examinations Board
		Wheatfield Road. Westerhop, Newcastle upon Tyne NE5 5JZ
	NWREB	North-West Regional Examinations Board
		Orbit House, Albert Street, Eccles, Manchester M30 OWL
	YHREB	Yorkshire and Humberside Regional Examinations Board
		Harrogate Office – 21–33 Springfield Avenue, Harrogate HG1 2HW
		Sheffield Office – Scarsdale House, 136 Derbyshire Lane, Sheffield S8 8SE
3 ▷ MIDLANDS EXAMINING GROUP (MEG)	Cambridge	University of Cambridge Local Examinations Syndicate
		Syndicate Buildings. 1 Hills Road, Cambridge CB1 2EU
	O & C	Oxford and Cambridge Schools Examinations Board
		10 Trumpington Street, Cambridge CB2 1QB, and Elsfield Way, Oxford OX2 8EP
	SUJB	Southern Universities' Joint Board for School Examinations
		Cotham Road, Bristol BS6 6DD
	WMEB	West Midlands Examinations Board
		Norfolk House. Smallbrook Queensway, Birmingham B5 4NJ
	EMREB	East Midlands Regional Examinations Board
		Robins Wood House, Robins Wood Road, Aspley, Nottingham NG8 3NR
4 ▷ SOUTHERN EXAMINING GROUP (SEG)	AEB	The Associated Examining Board
		Stag Hill House, Guildford, Surrey GU2 5XJ
	Oxford	Oxford Delegacy of Local Examinations
		Ewert Place, Summertown, Oxford OX2 7BZ
	SREB	Southern Regional Examinations Board
		Avondale House, 33 Carlton Crescent, Southampton, S_9 4YL
	SEREB	South-East Regional Examinations Board
		Beloe House, 2–10 Mount Ephraim Road, Tonbridge TN1 1EU
	SWEB	South-Western Examinations Board
		23–29 Marsh Street, Bristol BS1 4BP
5 ▷ WALES	WJEC	Welsh Joint Education Commitee
		245 Western Avenue, Cardiff CF5 2YX
6 ▷ NORTHERN IRELAND	NISEC	Northern Ireland Schools Examinations Council
		Beechill House, 42 Beechill Road, Belfast BT8 4RS
7 ▷ INTERNATIONAL GCSE	IGCSE	
		University of Cambridge Local Examiantions Syndicate
		1 Hills Road, Cambridge CB12 2EU

CHAPTER 2

COVERAGE OF TOPICS

This chapter considers the content of GCSE syllabuses in Mathematics at the Basic, Intermediate and Higher Level examinations respectively. It helps you see how the chapters and topics covered in this book relate to *each of these levels*. You can also check the content of this book with the examination set by your own particular Board (or Group).

BASIC
INTERMEDIATE
HIGHER

B A S I C L E V E L

The GCSE syllabuses at the *basic level* are almost identical for every Examination Board. **Every part** of this book is relevant to the basic level examination and assessment.

This *basic level* is referred to in a slightly different way by the various examination boards:

London & East Anglian Group (LEAG)	– Level X
Northern Examining Association (NEA)	– Level P
Midland Examining Group (MEG)	– Foundation
Southern Examining Group (SEG)	– Level 1
Welsh Joint Education Committee (WJEC)	– Level 1
Northern Ireland Schools Examinations Council (NISEC)	– Basic
Cambridge University Examinations Syndicate (IGCSE)	– Core

1 ▷ **CHAPTER AND TOPIC**

4	**Pattern in number**	whole numbers, odd, even, prime, multiples, factors, simple sequences
	Fractions	vulgar (simple addition and subtraction), decimal (the four rules on simple cases), percentage, equivalence, conversion of vulgar to decimal
	Directed number	in practical situations
	Square root	of perfect squares
5	**Approximation**	of length and weight of everyday objects
	Rounding off	to nearest 10p, 100, etc.
	General units	100 cm = 1 m, etc., and when to use them
	Time	12/24-hour clock, timetables
6	**Household finance**	HP, interest, taxation, discount, loans, wages and salaries, profit and loss, VAT
	Use of tables and charts	tidetables, conversion tables, insurance tables, etc.
	Reading scales	clocks, dials, scales
7	**Simple ratio**	sharing, recipes, scale drawing
	Proportion	direct and inverse
	Rates	speed, foreign currency exchange rates
8	**Formula**	flowcharts and substitution
	Brackets	an order of priority
9	**Cartesian co-ordinates**	plotting points and joining up
	Graphs	drawing graphs from given data
	Interpretation	of given graphs, travel graphs and conversion graphs
10	**Angles**	names and in triangles
	Plane figures	triangles, quadrilaterals, circles, similarity
	Symmetry	line, rotational
	Solid figures	names of, nets
11	**Perimeter**	of circles and rectilinear shapes
	Area	rectangle, triangle
	Volume	of cuboid
12	**Equipment**	protractor, compasses, set square and ruler
	Construction	of triangle, rectangle, and circle
	Drawings	accurate and scale drawings
	Bearings	compass points
13	**Transformation geometry**	tessellations, reflections and simple enlargements; rotations of 90°, 180°

14	Charts	interpretation of bar chart, pictogram and pie charts
	Frequency distribution	construction of bar chart, pictogram and tally chart
	Average	mode, median and mean
	Probability	simple equally likely situations

I N T E R M E D I A T E   L E V E L

The *intermediate levels* of all GCSE syllabuses are similar, although there are some differences. The table below will indicate the parts of this book which are relevant to each particular intermediate syllabus. Much of the book is relevant to most syllabuses.

The *intermediate level* is referred to by the different boards as:

LEAG– Level Y
NEA – Level Q
MEG– Intermediate
SEG – Level 2
WJEC– Level 2
NISEC– Intermediate
IGCSE– Core

Each syllabus also includes *all of the basic level* content.

CHAPTER AND TOPIC

	LEAG			NEA			MEG				SEG	WJEC	NISEC	IGCSE
	A	B	SMP	A	B	C	Maths	Mature	SMP(11–16)	SMP				
4 Integers, irrationals, prime factors, sequences	√	√	√	√	√	√	√	√	√	√	√	√	√	√
Fractions: conversion between fractions and percentage, standard form	√	√	√	√	√	√	√	√	√	√	√	√	√	√
Directed number: the 4 rules	√	√	√	√	√	√	√	√	√	√	√	√	√	√
Sets and Venn diagrams				√	√									
5 Rounding off: significant figures, decimal places	√	√	√	√	√	√	√	√	√	√	√	√	√	√
6 Simple interest and compound interest				√	√	√						√	√	
7 Ratio: scale factors, best buys	√	√	√	√	√	√	√	√	√	√	√	√	√	√
8 Formulae: transposition of	√	√	√	√	√	√	√	√	√	√	√	√	√	√
Algebraic factors, simplification, quadratic expansion	√	√	√	√	√	√	√	√	√	√	√	√	√	√
Equations: linear, inequalities	√	√	√	√	√	√	√	√	√	√	√	√	√	√
Indices: integral (positive and negative)	√	√	√	√	√	√	√	√	√	√	√	√	√	√
9 Constructing tables and graphs from equations	√	√	√	√	√	√	√	√	√	√	√	√	√	√
Gradient	√	√	√	√	√	√	√	√	√	√	√	√		√
Solution of simultaneous equations by graph	√	√		√	√	√	√	√	√					

| | LEAG | | | NEA | | | MEG | | | | SEG | WJEC | NISEC | IGCSE |
	A	B	SMP	A	B	C	Maths	Mature	SMP(11–16)	SMP				
10 Angles in parallels, polygons	✓	✓	✓	✓	✓	✓	✓	✓	✓	✓	✓	✓	✓	✓
Plane figures: angles in semi-circle, tangents	✓	✓	✓	✓	✓	✓	✓	✓	✓	✓	✓	✓	✓	✓
Congruency	✓	✓	✓	✓		✓	✓	✓	✓	✓	✓	✓	✓	
Angles in same segment, angles at centre of circle		✓												
11 Length of arc, area of sector				✓	✓	✓							✓	
Area of circle, parallelogram	✓	✓	✓	✓	✓	✓	✓	✓	✓	✓	✓	✓	✓	✓
Area of trapezium	✓	✓	✓				✓	✓				✓		✓
Volume of cylinder	✓	✓	✓	✓	✓	✓	✓	✓	✓	✓	✓	✓	✓	✓
Volume of prism	✓	✓					✓	✓	✓			✓		✓
Pythagoras and trigonometry	✓	✓	✓	✓	✓	✓	✓	✓	✓	✓	✓	✓	✓	✓
Plans and elevations								✓						
12 Drawing quadrilaterals, bisectors, constructing angles	✓	✓	✓	✓	✓	✓			✓	✓	✓	✓	✓	
Bearings: one point from another	✓	✓	✓	✓	✓	✓	✓	✓	✓	✓	✓			✓
13 Transformations: reflections			✓	✓	✓	✓	✓		✓	✓	✓			✓
rotations of 90°, 180°	✓	✓	✓	✓	✓	✓	✓		✓	✓	✓			✓
translations	✓	✓	✓	✓	✓	✓	✓		✓	✓	✓			✓
Vectors: graphical representation				✓	✓				✓					
14 Pie chart: construction of	✓	✓	✓	✓	✓	✓			✓	✓	✓	✓	✓	✓
Histogram: with equal interval	✓	✓	✓	✓	✓	✓	✓	✓		✓	✓	✓	✓	✓
Frequency distribution, grouped data				✓	✓	✓	✓	✓						
Scatter diagrams				✓		✓	✓	✓						
Probability: combined events	✓	✓	✓	✓	✓	✓	✓	✓	✓	✓	✓	✓	✓	✓

H I G H E R   L E V E L

The highest levels of the GCSE syllabuses now start to differ far more, so you do need to take note of the differences and to think how they will affect you. The table below will guide you on which parts of the book are relevant to your higher level syllabus.

The *higher level* is referred to by the different boards as:

LEAG– Level Z
NEA – Level R
MEG– Higher
SEG – Level 3
WJEC– Level 3
NISEC– High
IGCSE– Extended

Each syllabus also includes all the basic level content and the corresponding content at the intermediate level.

3 ▶ CHAPTER AND TOPIC

Chapter and Topic	LEAG A	LEAG B	LEAG SMP	NEA A	NEA B	NEA C	MEG Maths	MEG Mature	MEG SMP(11–16)	MEG SMP	SEG	WJEC	NISEC	IGCSE
4 Sets and Venn diagrams	√	√		√	√	√					√	√		√
6 Compound interest				√	√	√	√	√					√	
7 Ratios of similar shapes and volumes	√	√		√	√	√	√	√	√	√	√	√	√	√
Direct and inverse variation (proportion)	√	√	√	√	√	√		√	√	√	√	√	√	√
8 Quadratic factorisation, equations	√	√	√	√	√	√	√	√	√	√	√	√	√	√
Simultaneous equations	√	√		√	√	√	√	√	√	√	√	√	√	√
Algebraic fractions						√	√	√	√					
Fractional indices				√			√	√	√		√	√	√	
Functions, combinations			√	√	√	√			√	√	√	√		
9 Drawing graphs of inequalities				√	√		√		√	√	√		√	
Drawing graphs of simultaneous equations	√	√	√	√	√	√	√	√	√	√	√	√		√
Area under a graph	√	√		√	√	√	√	√			√	√		√
Gradient at a point and interpretation	√	√	√	√	√	√	√	√			√	√	√	√
Cyclic quadrilaterals	√	√		√							√	√		
10 Angles in a circle from chord, double centre	√	√	√	√							√	√	√	√
Axes and planes of symmetry	√	√		√							√			
Intersecting chord (internal)	√	√												
11 Length of arc, area of sector	√	√		√	√	√	√	√	√	√	√		√	√
Area of trapezium	√	√	√	√	√	√	√	√	√	√	√	√		√
Area using sine rule ($\frac{1}{2}ab\sin C$)							√	√				√		√
Volume of prism	√	√		√	√	√	√	√	√	√	√	√	√	√
Volume of sphere, cone	√	√		√	√	√	√	√	√	√	√	√	√	√
Surface areas	√	√		√	√		√	√	√	√	√	√	√	√
3D solution using trigonometry and Pythagoras	√	√	√	√	√	√	√	√	√	√	√	√		√
Sine rule							√	√				√		√
Cosine rule							√	√				√		√
12 Constructing a perpendicular, loci				√										
13 Vectors	√	√	√	√	√	√	√	√	√	√	√	√		√
Matrices	√	√	√	√	√		√		√		√			√
Transformations: combinations of	√	√	√	√	√		√		√		√		√	√
Enlargements with negative or fraction scale factor	√	√	√	√	√		√				√		√	√

	LEAG			NEA			MEG				SEG	WJEC	NISEC	IGCSE
	A	B	SMP	A	B	C	Maths	Mature	SMP(11–16)	SMP				
14 Cumulative frequency, quartiles				✓	✓	✓	✓	✓	✓	✓			✓	✓
Unequal interval histogram				✓		✓		✓						✓
Probability (and/or)	✓	✓		✓	✓	✓			✓	✓	✓		✓	

THE EXAMINATIONS

It is helpful to remember that if you have been correctly entered (and your teachers will usually be very good at this), then you can do at least half the examination questions well. This should give you a lot of confidence before you go into the examination. Being confident is helpful since, when you are anxious, you tend to make careless mistakes.

EXAMINATION AND ASSESSMENT TECHNIQUES

C A L C U L A T O R S

All GCSE examinations allow you to have your calculator available. The questions will be set on the assumption that you have a calculator suitable for your level. So, if you know you will be asked some trigonometry questions at your level, make sure you have the appropriate calculator. If you are being entered for the *basic* level then you will not need a scientific calculator, only a *standard* one. However, both *intermediate* and *higher* levels will require you to have a *scientific calculator* (or to use trigonometry tables and a standard calculator). It is up to **YOU** to be responsible for your calculator and not the examination board, school or college. Do have the right one, and make certain that the batteries are not going to run out (perhaps take some spares). Do use a calculator that you are familiar with, and not a strange one borrowed at the last minute.

When using the calculator in the examination, do not forget to write out your *method of solution*, otherwise you will often lose marks. In marking exam papers this year the answer to one question should have been £1.99. Some candidates gave the answer as £1.98 *but showed no working* hence they got no marks at all even though it is quite likely that they knew what they were doing but had just made a small error, perhaps in rounding off. You will throw marks away if you do not put down your method of solution.

F O R M U L A E   L I S T

Each Examination Board will supply a *formulae list* for each syllabus, and for each level in that syllabus. You are advised to become familiar with this list, so that you know where to find the formulae when needed. It is also important that you practise using those formulae. If you have practised using the formulae *before* the examination then this will give you confidence in using them in the examination itself.

R E V I S I O N

There is, of course, no substitute for hard work *throughout* the course, and for regularly doing homework and classwork assignments. Revision is, however, important and should be started well before the examination, best of all *before* the Easter holiday leading to the examination. The best way to revise mathematics is to *do* it. You should try as many questions as you can beforehand; this is why there are a lot of questions at the end of each chapter. Do not be afraid of going through the same question more than once during your revision. This will be helpful practice in using the correct technique for answering that type of question, and it should help boost your confidence. Do not revise for too long at a single sitting! You are advised to revise in short periods of between 45 to 60 minutes then to have a break before doing any more. Of course, this will vary with individuals, but if you've started your revision early enough this is usually the best way rather than a last final fling!

Use this book to remind you of the things you have been taught. Go through the worked examples then try the exercises for yourself, checking the answer before going any further. Finally, try the examination questions at the end of each chapter, making sure that you put down all your working, just as you will have to do in the examination.

EXAMINATION ROOM STRATEGY

Remember, you can do at least half the questions, and there will always be some that cause problems. You must use your time properly, so do not waste it. The majority of GCSE examinations use 'Question and Answer Books', which means there is space for you to work out your answer and to give an answer on the examination paper itself. So it doesn't matter in what order you do the questions. Go through the paper and *answer the questions you can do first*, then go back and attempt the ones you've left out. If a question causes you particular problems and you cannot see what to do then leave it, go on to another, and come back to it later. In other words, 'do what you can do well' first. This will help you 'put marks into the bank' and will help you gain confidence before you tackle the more 'difficult' questions.

Most examination papers will tell you how many marks are available to a question; the *more difficult* a question is, the *more marks* are generally given to it. So if you come across a question worth 5 marks and one worth 2 marks, you should expect the 2-mark question to be answered more easily than the 5-mark question. If you have managed to do the 5-mark question very easily, much more easily than the 2-mark question, just check that you have in fact done the question that has been set and not misread it! If you're answering on an answer booklet do also use *the number of lines* left for your answer as a guide. If only one line is given for working, then you should not need to do a lot of working. If, however, five lines have been given for working, then you should expect to need to complete a number of stages to get to the answer.

The number of marks per question will also give you some idea of how much time to spend on each question. Suppose the examination paper lasts $2\frac{1}{2}$ hours (150 minutes) and there are 100 marks, then each mark has an average time of $1\frac{1}{2}$ minutes. A 5-mark question should not take more than 8 minutes. Of course you should perhaps allow 10 minutes at the start of the examination for reading through the paper (or booklet) carefully and choosing your early questions, and perhaps 10 minutes for checking at the end. In this case you would be able to use the 'rule of thumb' that you have just over one minute per mark. Working out the *minutes per mark* should not be taken *too* far, but it does give you some idea of how to use your time well in the examination.

Finally, do not forget to *check* those answers, especially the sense and the accuracy of your answers. If you have calculated the cost of a car to be £6, you ought to suspect that your answer is wrong and check it. Year after year examiners always mark papers where 'stupid' answers are given, such as a man being paid a salary of £45 a year! Do check your answers, it will gain you marks. Also, check that you have *rounded off* suitably. Many questions will say 'round your answer to 1 decimal place', etc., in which case you could obtain marks for rounding off. But other questions (especially at the intermediate and higher levels) might simply say 'calculate the distance ...', and if your answer is something like 8.273419 km, you are quite likely to lose a mark for your answer since it is not given to a suitable degree of accuracy. Use the guidelines indicated in Chapter 5 to round off, or be prepared to lose marks.

You ought to be doing many of these checks while answering the question the first time, but do go through the routine as a check at the end. It may at the end be boring, but if it gains you those marks you would otherwise have lost and this makes the difference between grades, it will have been well worth doing.

EXAMINATION EQUIPMENT

You will be required to *calculate, draw* and *construct*. You must therefore have the right equipment for the job. Do not rely on the school providing it, since if you provide the equipment you are familiar with, you can be more confident that you can use it and rely on it. Make certain you have the following:

▶ calculator
▶ batteries for calculator
▶ ruler
▶ sharp pencils
▶ pen (and a spare pen)

▶ pencil sharpener
▶ rubber
▶ protractor (angle measurer)
▶ pair of compasses
▶ set square

EXAMINATION QUESTIONS

There are different *types* of question that you could meet: e.g. multiple choice, short answer, structured or combination.

A multiple-choice question is a short question with four or five different answers from which to choose the right one. The way to answer these questions is to actually *do* the question and then see if your answer matches one of those given. If not, then you know that you've gone wrong and can look again. Do not be tempted to guess at the most obvious answer straight away but do work it out. If at the end of the examination you find there are some of these multiple-choice questions not done then it is legitimate to have an intelligent guess at the answers (but this should always be the last resort). Only the LEAG intend to use this type of question at the moment and this is only part of their Paper 3.

Example 1   *Given that $Y = 2x^2 - 3x - 7$, what is the value of Y when $x = 3$?*
(A) 16;   (B) 9;   (C) 8;   (D) 2;   (E) −10

Here you would do the question first by substituting into the equation $x = 3$ then checking that your answer of 2 is one of the answers given.

This type of question is usually given one or two marks and you may only have a line or two on which to answer the question. You will be able to assess what you have to do, then be sure to write down the *method* you are using as well as the *answer*, suitably *rounded off* if necessary.

Example 2   *A Rock Group from America came to Britain on tour and went home with £14 000 profit made on all the concerts. The exchange rate was £1 to $1.12 at the time they set off from Britain. How many dollars would this make their profit?*

Here you need to sort out that it is necessary to multiply the £14 000 by 1.12 which is best done on the calculator to give $15 680. Your answer should include the statement $14 000 \times 1.12$ as the method you have used. You take the risk of losing marks here if you give only the answer without the statement.

These are the longer questions that will use one answer as part of the *next question*, perhaps two or three times in the one question. It is also vital that you show all your *method of working* here as one wrong answer early on will make all subsequent answers wrong and to gain your marks you must now show exactly what you have done.

Example 3   *John and his wife Mary wanted a new carpet measuring 4 m by 5 m for their lounge. They chose an Axminster carpet priced at £12.75 per square metre, with an extra charge for fitting of £1.75 per square metre. John and Mary agreed hire purchase terms with the shop of 10% deposit and the remainder to be paid in 12 equal monthly amounts.*

(a) *What is the area of the lounge floor to be carpeted?*
(b) *What would be the total cost of a fitted carpet for the lounge?*
(c) *Find what deposit is paid on the carpet.*
(d) *Calculate the monthly payment on the carpet.*
(e) *After a while, Mary wanted to clean the carpet with 'Kleenit'. A packet of 'Kleenit' is sufficient to clean 6 m² of carpet. How many packets should Mary buy in order to clean the whole carpet?*

You can see how one answer leads on to the next and that any mistakes made earlier will make a wrong answer appear later ... hence it is vital that you show all your method of solution in *each section* of the question. If you did this problem correctly then you would find that your answer to part (d) was £21.75.

| 4 > | COMBINATION QUESTIONS |

A longer question can, and often is, a *combination* of short-answer and structured questions.

Example 4   *A Girl Guide leader had to share a full cylindrical tin of cocoa into small cuboid boxes for her Girl Guides to use. The large cylindrical tin was 15 cm tall and had a diameter of 10 cm. Each cuboid box measured 4 cm by 4 cm by 3 cm.*

(a) *Calculate the volume of the cylindrical cocoa tin.*
(b) *Calculate the volume of one small cuboid box.*
(c) *How many boxes can she fill from the full cocoa tin?*

You should set out the first two parts as short questions with your method clearly stated. Then the final part is done by dividing the answer to (a) by the answer to (b). This must be very clearly written down since it is possible for you to have made a mistake in one of the two parts, and the examiner marking your paper needs to be able to see what you've done and not to have to do *your* sum himself just to check what you've done.

SUMMARY
To summarise this section we can simply say that at all stages you should show the method of solution, unless you are certain that there is only one mark for the question and that no method is being looked for.

C O U R S E W O R K

It is the intention that coursework should be an important part of the assessment (see Chapter 2), so it is important that your assignments be well planned throughout the course rather than become an unwelcome burden at the end of the course.

Your school or college will be responsible for deciding upon the actual nature of the coursework and the way in which it is organised. There are many different tasks you could be asked to do. Whatever the nature of the coursework, the assessment will be made in three main areas: *Practical, Investigation* and *Extended pieces of work*.

5 > PRACTICAL WORK

You will be assessed on:

(a) how you *planned the task*, how you carried it out and how accurate you were (evidence of these three stages is necessary);

(b) whether you have demonstrated that you *understand the use that can be made of equipment*; for example, in weighing, that you have used an *appropriate* set of scales;

(c) your actual *skill in using* the equipment;

(d) your ability to *communicate* what you are doing – you could well be asked to explain why you did a certain thing, or why you used a piece of equipment in a particular way.

The tasks set will be at the level for which you are being considered. If you move up a level during the course then you should be given an opportunity of doing the practical work appropriate to this higher level.

6 > INVESTIGATION WORK

You will be assessed on:

(a) how you *planned* and *prepared* the set task;

(b) how much *relevant information* you were able to obtain and use;

(c) your ability to *communicate* what you have done – e.g. you could be asked to talk about the investigation as well as to write a clear solution;

(d) the extent to which you were able to *draw a valid conclusion*;

(e) how *far* you went with the investigation. Was it exhaustive?

Very often the same investigation will be set for all levels. It is up to you to demonstrate how well you have been able to pursue the investigation and to decide the point at which you stop.

EXAMPLE INVESTIGATION

Four straight lines all intersect each other. How many intersections will there be for other numbers of lines?

(a) This work can be planned in such a way that it can become an *investigation*. First you can draw two lines, then three, then four, and so on.

(b) You now need to look for a *pattern*. If you can identify a pattern you can start predicting how many intersections there will be for the next sets of lines without having to draw them.

(c) The work must be written up clearly. Start with an *introduction*, telling the assessor what you were trying to do. Follow this by outlining the *method* you have chosen to pursue your investigation. Present a *table of results*, giving an indication of any patterns you noticed.

(d) Can you draw a *conclusion*? For example, can you state how you can find the number of intersections for any given number of straight lines, say 50?

(e) How far have you been able to see a pattern? Can you write a *formula* for n lines, and the number of intersections this will give?

In an investigation it is up to you to go as far into the investigation as you can. But do be clear and logical in how you set about conducting the investigation. Make sure that you write up your results neatly and on the lines suggested: introduction → method → results → conclusion.

7 > EXTENDED WORK

Usually, the task set will be defined by the school or college from some particular starting point. It is then up to you where you take it and how far you develop it. The main points that will be looked for in an extended piece of work are given below.

(a) *The comprehension of the task*. Did you *understand the problem* and were you able to *define* what you were going to do?

(b) *Planning*. Were you able to *plan* out the task into different set stages to enable you to complete the task?

(c) *Peformance of the task*. How well did you *undertake the set task*. Did you choose appropriate methods? Dīd you use appropriate equipment? How have you interpreted the results from that equipment?

(d) *Communication*. This will be both written and oral. Again a well set-out introduction → method → results → conclusion will be important. Have you used helpful diagrams/tables, etc? In oral work were you able to respond to unexpected queries?

EXAMPLE PIECE OF WORK

Plan a return journey from your home to the nearest zoo (or country park) for yourself and some children, in one day.

That is a straightforward task, and the main points looked for on the assessment would be:

(a) Were you able to obtain bus/train timetables, opening hours of the zoo and prices?

(b) How well did you link the times together for setting off, lunch and returning?

(c) How appropriate is the actual solution? Does it give adequate time to get there, to enjoy the day, and return? How accurate are the prices, and how practical is the suggested journey? In fact, has the journey actually been made?

(d) How well is the task written up, and can you talk about it? In fact, if you actually undertook the task you could comment on the accuracy of the planning.

Your coursework tasks should be assessed at frequent, but appropriate, times during the course. Assessed coursework will help you to be aware of how well you are doing. If you do have shortcomings you can then work on improving them before the next time such material is assessed. Yet in every case, the coursework component comes down to *you*, inasmuch as it really does assess the way in which *you*

▶ plan the work
▶ do the work, and
▶ communicate the work.

It is all up to you!

NUMBER

You will be required to recognise the *patterns* that arise in numbers as a result of many different situations. You will also have to actually work out some calculations. In the vast majority of examinations you are allowed to use a *calculator*. Even so, you must be very careful to check your answers to see that they make sense. You should round your answers off to a suitable degree of accuracy, as you will see in the next chapter.

Many of the ideas in this chapter will, of course, be used in later chapters and will not necessarily be examined on their own.

USEFUL DEFINITIONS

Integer	a whole number
Fraction	a part of a whole
Equivalent	having the same value but looking different
Factors	an integer that exactly divides another integer
Multiple	the result of multiplying an integer by an integer
Prime	a number that has two factors only
Sequence	a list of numbers that follow a pattern or rule
Square	to multiply a number by itself
Set	a collection of elements that have something in common
Venn diagram	a pictorial way of illustrating sets
Finite	of known number
Infinite	when the number is too large to be countable

PATTERN IN NUMBER

NEGATIVE NUMBER

ARITHMETIC

FRACTIONS

PERCENTAGE

SPECIAL NUMBER NAMES

SETS AND VENN DIAGRAMS

E S S E N T I A L   P R I N C I P L E S

PATTERN IN NUMBER

MULTIPLES

You know the *odd* numbers (1, 3, 5, 7, etc.) and the *even* numbers (2, 4, 6, 8, etc.). Even numbers are *multiples* of 2. In other words, 2 divides *exactly* into each *even* number. Other examples of multiples could be of 4 (e.g. 4, 12, ..., 96, 100, ...), of 5 (5, 10, 15, ..., 185, ...), or of any whole number.

FACTORS

The *factors* of a whole number N are the *whole numbers* that will *divide* into N exactly. So the factors of 12 are 1, 2, 3, 4, 6 and 12. The factors of 16 are 1, 2, 4, 8 and 16.

WORKED EXAMPLE 1

A sweet manufacturer wanted to put his chocolate bars into packs of more than 1 bar, and pack them into a box that can contain 100 bars of chocolate. What possible numbers can he put into a pack?

He can put any number into a pack as long as that number is a *factor* of 100. The factors of 100 are 1, 2, 4, 5, 10, 20, 25, 50 and 100. Since there must be more than 1, he has the choice of any of the others.

EXERCISE 1

Find the factors of 72.

PRIME NUMBERS

A *prime number* is a whole number that has *two*, and only two, *factors*. For example, 7 has only two factors: 1 and 7, and 13 also has only two factors: 1 and 13. 1 is not a prime number as it has only one factor (1) and not two. Try to write down the first ten prime numbers (2 is the first, and 29 is the tenth).

WORKED EXAMPLE 2

In one school year there were four classes. Class M4A had 29 pupils, Class M4B had 27 pupils, Class M4C had 23 pupils and Class M4D had 22 pupils. Two of the teachers said they always had problems when trying to put the children into equal groups. Which classes did these teachers take and how could the Headmaster have avoided the problem?

The classes with 29 and 23 pupils cannot be divided into equal groups because they are prime numbers. So the teachers from M4A and M4C would have complained. To avoid the problem, the Head could change the numbers in each class so that none contains a prime number.

PRIME FACTORS

The *prime factors* of an integer (whole number) are factors that are prime numbers. For example, the prime factors of 35 are 5 and 7, which, for convenience, we usually write as 5×7, and the prime factors of 12 are $2 \times 2 \times 3$ (note how we put the 2 down twice so that the product of these factors gives the integer we start with). You can check for yourself that the prime factors of 72 are $2 \times 2 \times 2 \times 3 \times 3$, which we would shorten to $2^3 \times 3^2$, and that the prime factors of 90 are $2 \times 3^2 \times 5$.

EXERCISE 2

Find the prime factors of (i) 100 and (ii) 130.

COMMON FACTORS AND HCF

Common factors are the factors that two integers have in common. For example, the common factors of 16 and 24 are 2, 4 and 8, since all three numbers divide exactly into 16 and 24. Here 8 is the HCF, since it is the highest common factor.

One way to find the HCF is to consider the prime factors. For example, to find the HCF of 72 and 90, we break each number into its prime factors, $2 \times 2 \times 2 \times 3 \times 3$ and $2 \times 3 \times 3 \times 5$, then look for the figures each has in common. Here both have in common $(2 \times 3 \times 3)$, hence the HCF is 18.

EXERCISE 3

Find the HCF of 36, 90 and 108.

COMMON MULTIPLES AND LCM

Common multiples are the multiples that two integers have in common. For example, the common multiples of 6 and 8 are 24, 48, 72, . . . Here 24 is the LCM, since it is the lowest common multiple.

Again, we can use prime factors to help us find the lowest common multiple. For example, to find the LCM of 75 and 90 we break each number into its prime factors, $3 \times 5 \times 5$ and $2 \times 3 \times 3 \times 5$, then look for the smallest combination of both, which here is $2 \times 3 \times 3 \times 5 \times 5$. (Notice you can find both previous prime factors here.) Hence the LCM of 75 and 90 is 450.

EXERCISE 4

Find the LCM of 18 and 30.

SEQUENCES

Sequences are lists of numbers that follow a pattern. For example, in the sequence 3, 7, 11, 15, 19, . . you add 4 each time. In the sequence 3, 6, 12, 24, . . . you double the figure each time, and in the sequence 1, 1, 2, 3, 5, 8, 13, 21, . . . you add the last two terms to get the next in the sequence.

WORKED EXAMPLE 3

Find the next three numbers in the sequence 1, 2, 4, 7, 11, 16, ...

Look at how the pattern builds up. You add on 1, then 2, then 3, then 4, etc., so the next three numbers will be 22, 29 and 37.

SQUARE NUMBERS

A *square number* is a number that can be formed by multiplying a whole number by itself. For example, 25 is a square number because 5 multiplied by itself is 25. Try to write down the first ten square numbers (1 is the first and 100 is the tenth).

SQUARE ROOTS

Square roots are numbers which, when multiplied by themselves, give you a particular number. For example, the square root of 36 is 6, since 6×6 is 36. We use a special mathematical sign for square root, it is $\sqrt{}$ (find it on your calculator). So $\sqrt{9}$ means the 'the square root of 9'. Now, on your calculator press 9, followed by $\sqrt{}$ and you should get 3, so $\sqrt{9} = 3$.

EXERCISE 5

Find both the square and the square root of (i) 16 and (ii) 0.9

NEGATIVE NUMBERS

We use negative numbers most often in winter when we talk about the temperature, since temperatures below freezing point are 'negative numbers'.

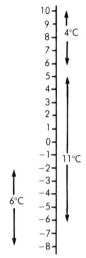

Look at this thermometer. It shows a reading of 7° below freezing point. We call this minus 7 °C and write it as −7 °C.

To work out the difference between two temperatures we can consider a temperature scale such as the one alongside.

You can see that the differences between

(i)   10 °C and   6 °C is 10 − 6
      which is 4 °C
(ii) 5 °C and −6 °C is 5 + 6
      which is 11 °C
(iii) −2 °C and −8 °C is 8 − 2
      which is 6 °C

It is necessary to be able to calculate new temperatures when we are given a rise or fall from a previous temperature. For example, if the temperature is 8 °C and falls by 10 °C, then by counting down 10 °C from 8 °C we come to −2 °C which is the new temperature. Also, if the temperature is −10 °C and rises by 4 °C, then by counting up 4 °C from −10 °C we come to the new temperature of −6 °C.

> **2**   **NEGATIVE NUMBER ARITHMETIC**

You need to be able to add, subtract, multiply and divide with negative numbers.

ADDING AND SUBTRACTING

There are some little *rules* that you ought to know:

> **+ − is the same as −**
> **− − is the same as +**

> 66 This is where a lot of errors are made so read this part carefully 99

So, for example, 5 + −2 is the same as 5 − 2, which is 3

and

5 − −2 is the same as 5 + 2, which is 7.

You can see from the thermometer that by going up and down the temperature scale you get:

$$3 + 8 = 11, \quad 3 − 8 = −5, \quad −3 + 8 = 5, \quad −3 − 8 = −11.$$

In the same way,

$$3 + −8 = 3 − 8 = −5 \quad \text{and} \quad 3 − −8 = 3 + 8 = 11.$$

The Figure illustrates how we calculate −3 + 5 to be 2 or −3 − −5 to be 2.

(a) The first thing to do is to start with the first number on the number scale: here it is −3.

(b) Since the second number is '+' or '− −' (+5 or − −5 in this case), count 5 numbers up the scale and you see that you come to 2.

Note when the second number is '−' or '+ −' you go down the scale.

MULTIPLYING AND DIVIDING

Here are some more *rules* you ought to know: (The same rules apply when dividing.)

Rules for multiplying

+ × + = +	
− × − = +	
+ × − = −	
− × + = −	

Signs the same; answer: +

Signs different; answer: −

That is: when the signs are the *same*, the answer is +
      when the signs are *different*, the answer is −

For example,

$$5 \times \ \ 3 = 15 \qquad 2 \times -3 = -6 \qquad 8 \div \ \ 4 = 2 \qquad 6 \div -2 = -3$$
$$-5 \times -3 = 15 \qquad -2 \times \ \ 3 = -6 \qquad -8 \div -4 = 2 \qquad -6 \div \ \ 2 = -3$$

This will be of use to you when substituting into given formulae in later chapters.

EXERCISE 6

Calculate: (i) $-3 + 5$; (ii) $-3 - 5$; (iii) -3×-5; (iv) -3×5; (v) $15 \div -3$.

---

3 > FRACTIONS

There are two types of fractions you should be familiar with: *vulgar fractions* and *decimal fractions*.

VULGAR FRACTIONS

A vulgar fraction is always expressed using two numbers, one above the other, for example $\frac{3}{5}$.

The shaded region of the rectangle represents $\frac{3}{10}$ (or three tenths). The 10 comes from the fact that the rectangle is divided into 10 equal pieces and the 3 because 3 pieces are shaded. We use this idea when finding a fraction of a given amount.

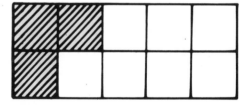

WORKED EXAMPLE 4

Find $\frac{2}{5}$ of £8.

First divide £8 by 5 to find $\frac{1}{5}$, which will be £1.60. Now multiply this by 2 to find $\frac{2}{5}$, giving a total of £3.20.

EQUIVALENT FRACTIONS

Many vulgar fractions are the same as each other, or, as we would say, are *equivalent*. These diagrams illustrate some fractions all equivalent to $\frac{3}{4}$.

$$\frac{3}{4} \qquad\qquad \frac{6}{8} \qquad\qquad \frac{9}{12} \qquad\qquad \frac{12}{16}$$

You can find a lot more equivalent fractions just by multiplying the top part of the fraction and the bottom part by the same number.

We use this idea in what we call *cancelling down* to the simplest equivalent fraction. You may use other phrases for this, like 'simplifying' or 'putting into lowest terms'. We do this by dividing the top and the bottom by the same number. For example, $\frac{12}{20}$ will simplify down to $\frac{3}{5}$, since we can divide both the top and the bottom by 4.

WORKED EXAMPLE 5

Mat has just lost 24 sheep from his flock of 80 through foot and mouth disease. What fraction of his flock, written as simply as possible, has he lost?

The fraction lost is 24 out of 80, or $\frac{24}{80}$. Both top and bottom can be divided by 8, so we can cancel using 8 to give us $\frac{3}{10}$.

ADDITION AND SUBTRACTION

You need to be able to add and subtract simple *vulgar* fractions. Remember that it is only easy to do this when the bottom numbers are the same. For example, $\frac{3}{10}$ added to $\frac{4}{10}$ is $\frac{7}{10}$, or $\frac{3}{5} - \frac{1}{5} = \frac{2}{5}$. So if the bottom numbers are *not* the same then you need to change them by using equivalent fractions. For example, if we want to add $\frac{1}{2}$ and $\frac{3}{8}$, we need to make both bottom numbers the same; we can make both of them 8 by multiplying, hence $\frac{1}{2}$ becomes $\frac{4}{8}$ while $\frac{3}{8}$ stays as it is. This gives us $\frac{4}{8} + \frac{3}{8}$ which is $\frac{7}{8}$.

WORKED EXAMPLE 6

Kevin sold $\frac{1}{4}$ of his stamp collection to Brian, and gave $\frac{3}{8}$ of it away to Malcolm. How much of his collection had he left?

We need to add together $\frac{1}{4}$ and $\frac{3}{8}$. We can change $\frac{1}{4}$ to $\frac{2}{8}$ by thinking about the equivalent fractions, then add $\frac{2}{8}$ and $\frac{3}{8}$ to give $\frac{5}{8}$. Now we think about $1 - \frac{5}{8}$ to find out what fraction is left. The 1 can be written as $\frac{8}{8}$, hence the sum becomes $\frac{8}{8} - \frac{5}{8}$, which is $\frac{3}{8}$.

MIXED NUMBERS

When adding two vulgar fractions, say $\frac{7}{8}$ and $\frac{5}{8}$, we get $\frac{12}{8}$ which is 'top heavy'. In other words, more than a simple fraction with one or more 'whole numbers' involved. 'Top heavy' fractions can be simplified to what we call *mixed numbers*.

WORKED EXAMPLE 7

Ross bought 7 bottles of ginger beer, each containing $\frac{3}{4}$ pint. How much had he altogether?

We need 7 lots of $\frac{3}{4}$ or $7 \times \frac{3}{4}$ which is $\frac{21}{4}$. As four quarters make a whole item we divide 21 by 4 to get 5 remainder 1, which would be 5 whole items and 1 quarter. So Ross had $5\frac{1}{4}$ pints altogether.

FURTHER ADDITION AND SUBTRACTION

You must also be able to add two fractions like $\frac{2}{5}$ and $\frac{1}{3}$ where we need to change both fractions to get the same bottom number. Look for the LCM of 5 and 3, which is 15, and make both fractions into fifteenths. We can illustrate this by:

$$\frac{2}{5} + \frac{1}{3} \rightarrow \frac{6}{15} + \frac{5}{15} = \frac{11}{15}.$$

In a similar way we can illustrate $\frac{5}{6} + \frac{1}{8}$ where 24 is the LCM of the bottom numbers:

$$\frac{5}{6} + \frac{1}{8} \rightarrow \frac{20}{24} + \frac{3}{24} = \frac{23}{24}.$$

If one or both of the numbers being added is a mixed number, then we add the whole ones separately then add the fractions, as illustrated in the next example:

$$2\frac{3}{4} + 1\frac{1}{2} \rightarrow 3 + \frac{3}{4} + \frac{1}{2} \rightarrow 3 + \frac{5}{4} \rightarrow 3 + 1\frac{1}{4} = 4\frac{1}{4}.$$

EXERCISE 7

Calculate: (i) $\frac{3}{5} + \frac{1}{4}$; (ii) $3\frac{1}{2} + 2\frac{4}{5}$; (iii) $\frac{7}{8} - \frac{5}{6}$.

MULTIPLICATION OF FRACTIONS

You may well need at some time to *multiply* two fractions, for example $\frac{4}{5} \times \frac{1}{6}$. We do this by simplifying the question first by 'cancelling', to see if anything will divide into any of the top numbers and also any of the bottom numbers. Here we can 'cancel' both the 4 and the 6 by 2 to give the product $\frac{2}{5} \times \frac{1}{3}$ which we now multiply by multiplying the top numbers then multiplying the bottom numbers to give $\frac{2}{15}$. We can summarise this by:

$$\frac{4}{5} \times \frac{1}{6} \to \frac{{}^2\!\!\!\not4}{5} \times \frac{1}{\not6_3} \to \frac{2}{5} \times \frac{1}{3} \to \frac{2}{15}.$$

In a similar way we can illustrate $\frac{4}{9} \times \frac{3}{10}$, where more 'cancelling' can be done:

$$\frac{4}{9} \times \frac{3}{10} \to \frac{{}^2\!\!\!\not4}{9} \times \frac{3}{\not{10}_5} \to \frac{2}{\not9_3} \times \frac{\not3^1}{5} \to \frac{2}{3} \times \frac{1}{5} \to \frac{2}{15}.$$

You certainly would not be expected to write all this down as a solution. It has been written out in full here simply to illustrate what is done.

If one (or more) of the numbers is a mixed number then it needs changing to a 'top heavy' fraction before we can multiply. For example, $2\frac{1}{2} \times \frac{4}{5}$ can be shown as:

$$2\frac{1}{2} \times \frac{4}{5} \to \frac{5}{2} \times \frac{4}{5} \to \frac{{}^1\not5}{{}_1\not2} \times \frac{\not4^2}{\not5_1} \to \frac{2}{1} \to 2.$$

DIVISION OF FRACTIONS

To cut a long story short, you divide two fractions by turning the second one upside down and then multiplying.

WORKED EXAMPLE 8

Calculate: $\frac{4}{5} \div \frac{2}{3}$.
Change the $\frac{2}{3}$ to $\frac{3}{2}$ and multiply, hence

$$\frac{4}{5} \div \frac{2}{3} = \frac{4}{5} \times \frac{3}{2} = \frac{12}{10} = \frac{6}{5} = 1\frac{1}{5}.$$

DECIMAL FRACTION

The other type of fraction is a *decimal* fraction. This is shown by the numbers on the right-hand side of a decimal point. For example, 5.62 shows 5 whole ones and .62 is the decimal fraction. You need to be familiar with these equivalent fractions which help to show what decimal fractions are:

$$\frac{1}{10} = 0.1; \quad \frac{3}{10} = 0.3; \quad 4\frac{7}{10} = 4.7; \quad \frac{1}{100} = 0.01; \quad \frac{27}{100} = 0.27; \quad 26\frac{15}{100} = 26.15.$$

ADDING AND SUBTRACTING

You may say: 'But I have my calculator.' This may be true, but will you always have it wherever you are? I doubt it, so it is important that you know how to do this decimal arithmetic.

When adding or subtracting decimal numbers you must first line up your numbers so that the decimal points are underneath each other, then add or subtract, depending on the sum given.

WORKED EXAMPLE 9

A metal bar of usual length 23.4 cm expands by 1.76 cm when it is heated to 100 °C. What is its length at this temperature?

Add together 23.4 and 1.76 as 23.4
$$\underline{1.76}$$
$$\underline{25.16}\text{ cm}$$
(Try to do this first without a calculator.)

WORKED EXAMPLE 10

Gillian went into a shop with £15 and came out with £7.60. How much had she spent?

Subtract 7.60 from 15 as
$$\begin{array}{r} 15.00 \\ -\ 7.60 \\ \hline 7.40 \end{array}$$

(Note that it helps to change the 15 to 15.00) So Gillian had spent £7.40.

MULTIPLICATION

To multiply two decimal numbers (or more if you want to), you can follow this simple process.

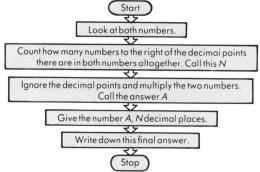

WORKED EXAMPLE 11

Find the cost of 0.4 kg of meat at £1.65 per kg.

You want to calculate 0.4×1.65; it has three decimal places. Ignore the decimal points and calculate 4×165 to give 660; give the answer to three decimal places which will be 0.660. The cost is £0.66.

WORKED EXAMPLE 12

Find the total length of three lengths of garden fencing each 1.6 metres long.

You want to calculate 3×1.6; it has just one decimal place. Ignore the decimal points and calculate 3×16 to give 48; give the answer to one decimal place, which will be 4.8 m.

DIVISION

You are only expected to be able to divide a decimal number by a whole number without a calculator. This is done in the same way as a 'normal' division, except that the decimal point of the answer will appear directly above the decimal point of the question.

WORKED EXAMPLE 13

How much will each child get if £21.50 is divided between five children?

Calculate $21.50 \div 5$ as $\quad 5\overline{)21.50}^{\ 4.30} \quad$ The answer is £4.30.

MENTAL CALCULATIONS

Naturally, wherever possible you would try to do these calculations on the calculator, but there are some very easy ones that can be done 'in the head' or 'mentally'. For example, when multiplying by 10 simply move the decimal point one place to the right as in the sums $5.67 \times 10 = 56.7$, and $3 \times 10 = 30$. When multiplying by 100 simply move the decimal point two places to the right, as in the sums $4.67 \times 100 = 467$, $0.597 \times 100 = 59.7$, $2.7 \times 100 = 270$ and $1.09 \times 100 = 109$.

A similar thing is true for dividing by 10, when we simply move the decimal point one place to the left, as in the sums $57 \div 10 = 5.7$ and $81.97 \div 10 = 8.197$. For dividing by 100, we move the decimal point two places to the left as in the sums $271 \div 100 = 2.71$, $25.9 \div 100 = 0.259$, $5 \div 100 = 0.05$ and $2.79 \div 100 = 0.0279$.

You can practise these decimal type calculations by setting yourself some similar problems, calculating them and then checking the answers on your calculator.

CONVERSION BETWEEN VULGAR AND DECIMAL

It is necessary for you to be able to convert vulgar fractions to decimal fractions. This helps you to compare fractions with each other and also to add or subtract awkward vulgar fractions.

To convert a vulgar fraction to a decimal fraction you just divide the top number by the bottom number (using a calculator if you have one available).

WORKED EXAMPLE 14

Which is the bigger fraction, $\frac{7}{8}$ or $\frac{17}{20}$?

Make each one into a decimal fraction: $\frac{7}{8}$ will become $7 \div 8$ which is 0.875, whilst $\frac{17}{20}$ becomes $17 \div 20$ which is 0.85, and since 0.875 is bigger than 0.85, then $\frac{7}{8}$ is bigger than $\frac{17}{20}$.

RECURRING DECIMALS

Both these last two fractions are what we call *terminating decimals*, because they have a fixed number of decimal places, but some fractions are not like that: they just seem to go on and on for ever. For example, try working out $\frac{1}{3}$. Do it on the calculator and you get 0.3333333, but do it the long way $\overset{0.333333\ldots}{3\overline{|1.000000\ldots}}$ and you will see it will go on and on for ever. We write this as $0.33\dot{3}$ – the dots mean this goes on and on for ever and we call these types of decimal numbers *recurring decimals*. Try $\frac{1}{11}$; do you get 0.09090909 ... ? We would write this as $0.09\dot{0}\dot{9}$. Note where the dots are this time, to show which figures repeat for ever. Try some more for yourself and make a list of the vulgar fractions that give terminating decimals and those that give recurring decimals.

4 ▷ PERCENTAGE

One per cent is written as 1%, which means 1 out of 100 or $\frac{1}{100}$ or 0.01. So 2% is the same as $\frac{2}{100}$ or 0.02, and 15% is the same as $\frac{15}{100}$ or 0.15. Using percentage with money can be simplified if we recognise that 1% of £1 is 1p and so 8% of £1 is 8p, etc. It follows then that, for example, 15% of £5 is $15 \times 1\%$ of £5, i.e. 15×5p which is 75p.

WORKED EXAMPLE 15

Joseph the paper boy, who earned £6 a week, was given a 12% pay increase. What is his new pay per week?

An increase means 'gets bigger', and 12% of £6 is 12×6p which is 72p. So Joseph's new pay is £6 + 72p, which is £6.72.

WORKED EXAMPLE 16

A shop reduced its prices by 10% in a sale. What was the new price of a radio that was previously marked £25?

A reduction means 'gets smaller', and 10% of £25 is 10×25p which is £2.50, so the new price of the radio is £25 − £2.50, which is £22.50.

EXERCISE 8
Find 15% of £5.60.

FRACTIONS INTO PERCENTAGES

To *change* a fraction to a percentage, simply multiply by 100. For instance, if Arun scored 18 out of 20 in a test, the percentage he would have got would be found by multiplying $\frac{18}{20}$ by 100. The easiest way to do this is on your calculator, as $(18 \times 100)/20$, which will come to 90%.

PERCENTAGE INCREASE

If we want to *increase* by, say, 5%, we really need to calculate 105% (100 + 5), which will simply mean multiplying by 1.05. For example, to increase £7 by 8% we can calculate £7 × 1.08, which is £7.56.

This is a far quicker way than finding 8% and adding on, but you can check that you get the same answer.

PERCENTAGE DECREASE

If we want to *decrease* by, say, 7%, we really need to calculate 93% (100 − 7) or multiply by 0.93. For example, £9.20 decreased by 20% is found by calculating £9.20 × 0.80, which is £7.36.

Again, check you get the same answer by the long method of finding 20% and subtracting.

EXERCISE 9

(i) Increase £50 by 6%;   (ii) decrease £800 by 17%.

---

5 ▷ SPECIAL NUMBER NAMES

You need to be familiar with various types of numbers.

INTEGER

Integers are 'whole numbers', either positive or negative, and zero; for example, $\ldots -4, -3, -2, -1, 0, 1, 2, 3, 4, \ldots$

RATIONAL NUMBERS

Rational numbers are all the numbers that can be expressed as a vulgar fraction of two integers. For example, 5 is a rational number, since 5 can be written as $\frac{5}{1}$. 9.16 is a rational number since it can be written as $\frac{916}{100}$. $\sqrt{9}$ is a rational number since it is 3 or −3.

It is worth remembering that all the *recurring decimals* you came across earlier came from fractions. For example, $\frac{2}{11}$ is $0.18\dot{1}\dot{8}$, $\frac{5}{7}$ is $0.\dot{7}1428\dot{5}$, and in fact all recurring decimals can be shown to be *rational* numbers.

IRRATIONAL NUMBERS

Irrational numbers are all numbers that cannot be expressed exactly as a vulgar fraction of two integers. For example, $\sqrt{2}$ cannot be expressed as a vulgar fraction, π cannot be expressed exactly as a fraction ($\frac{22}{7}$ is only the best approximation to it).

STANDARD FORM

Standard form is a convenient way of writing very large or very small numbers. It is always expressed in the terms of:

$$a \times 10^n,$$

where a is a number between 1 and 10 and n is an integer.

For example,   200   would be written as 2.0×10^2
and   617 000   would be written as 6.17×10^5
and   8431.3   would be written as 8.4313×10^3.

Notice how the number on the 10 (called the *index*) tells you how many places to move the decimal point. If the number is less than 1 to start with then we use negative indices on the 10. For example

0.015   will be 1.5×10^{-2}
0.000 000 783 will be 7.83×10^{-7}.

EXERCISE 10

Rewrite in standard form: (i) 568 900;   (ii) 0.000 527.

6 > **SETS AND VENN DIAGRAMS**

A *set* is a collection of 'things', usually with something in common. In mathematics this is shown by curly brackets, thus {...}. For example, {vowels} is the set a, e, i, o, u; which is a *finite* set because we know how many elements are in the set. An example of an *infinite* set is {even numbers} which is 2, 4, 6, ..., and we can never count them all.

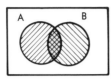 A = Auburn Hair
B = Brown Eyes

  Auburn Hair
and
Brown Eyes

A *Venn diagram* is a way of representing sets in a pictorial way.
The diagram shows the set of all girls with the two special sets of *A* and *B* as indicated. The region in the middle (criss-crossed) will be the auburn haired girls who also have brown eyes. The region left blank (outside both circles) will be the girls who have neither auburn hair nor brown eyes.

WORKED EXAMPLE 17

Of 10 people, 6 were working, 5 were married and 2 were neither married nor working. Find out how many people are married and working.

Put the information onto a Venn diagram.

We can see that (i) must contain 8 people, since 2 of the 10 are not included. The number of people in the centre part, (ii), is found by (6 + 5) − 8, which is 3. Hence, this centre part, representing the people who are both married and working, contains 3 people.

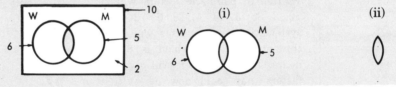

SET LANGUAGE AND NOTATION
Do check carefully whether your syllabus and level require this.
∩ means 'intersection'; that is, what is in both sets at the same time. For example,

$\{1, 2, 3, 4\} \cap \{2, 4, 6, 8, 10\} = \{2, 4\}$.

∪ means 'union'; that is, what is in both sets altogether. For example,

$\{5, 10, 15, 20\} \cup \{10, 20, 30\} = \{5, 10, 15, 20, 30\}$.

$\mathscr{E}$ means 'universal set'. This defines the limit of your situation.
∅ means { }, in other words, an empty set.
A' means 'the complement of *A*'; that is, what is not in *A*. This is why we need the $\mathscr{E}$. For example,

when $\mathscr{E} = \{1, 2, 3, 4, 5, 6, 7\}$ and $A = \{2, 4, 6\}$ then $A' = \{1, 3, 5, 7\}$.

n(A) means 'the number of *A*'; that is, how many elements are in *A*. For example,

if $A = \{2, 4, 6, 8, 10\}$ then $n(A) = 5$.

⊂ means a 'subset of' or 'is contained in'. For example

$\{2, 3\} \subset \{1, 2, 3\}$

∈ means 'is a member of'. For example

$6 \in \{2, 4, 6, 8\}$

Venn diagrams can be used to illustrate some of these ideas:

Shaded part represents $A \cup B$

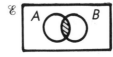
Shaded part represents $A \cap B$

Shaded part represents A'

Shaded part represents $(A \cup B \cup C)$

WORKED EXAMPLE 18

If $n(\mathscr{E}) = 40$ and both A and B are subsets of $\mathscr{E}$, where $n(A) = 30$, $n(B) = 21$ and $n(A \cup B)' = 2$, calculate $n(A \cap B)$.

Since $n(A \cup B)' = 2$, then $n(A \cup B) = 40 - 2 = 38$. So

$$n(A \cap B) = n(A) + n(B) - n(A \cup B)$$
$$= 30 + 21 - 38 = 13.$$

INEQUALITIES

You should be familiar with the four signs:

$>$	which means 'greater than'	e.g. $8 > 3$
$<$	which means 'less than'	e.g. $1 < 5.6$
$\geqslant$	which means 'greater than or equal to'	
$\leqslant$	which means 'less than or equal to'.	

SOLUTIONS TO EXERCISES

S1

Written in pairs the factors will be 1, 72; 2, 36; 3, 24; 4, 18; 6, 12; 8, 9. A systematic search in this way from 1, through to $\sqrt{72}$ which is 8 to the nearest whole number, will give all the factors that could now be written in order as:

1, 2, 3, 4, 6, 8, 9, 12, 18, 24, 36 and 72.

S2

(i) $100 = 2 \times 2 \times 5 \times 5 = 2^2 \times 5^2$;

(ii) $130 = 2 \times 5 \times 13$.

S3

$36 = 2 \times 2 \times 3 \times 3$; $90 = 2 \times 3 \times 3 \times 5$;

$108 = 2 \times 2 \times 3 \times 3 \times 3$;

hence HCF $= 2 \times 3 \times 3 = 18$.

S4

$18 = 2 \times 3 \times 3$; $30 = 2 \times 3 \times 5$;

hence, LCM $= 2 \times 3 \times 3 \times 5 = 90$.

S5

(i) square $= 16 \times 16 = 256$; $\sqrt{16} = 4$ and -4.

(ii) square $= 0.81$; $\sqrt{0.9} = 0.948\,683\,3$ (or a suitably rounded off answer).

S6

(i) 2; (ii) -8; (iii) 15; (iv) -15;

(v) -5.

S7

(i) $\dfrac{3}{5} + \dfrac{1}{4} \to \dfrac{12}{20} + \dfrac{5}{20} = \dfrac{17}{20}$.

(ii) $3\dfrac{1}{2} + 2\dfrac{4}{5} \to 5 + \dfrac{5}{10} + \dfrac{8}{10}$

$= 5 + \dfrac{13}{10} = 6\dfrac{3}{10}$.

(iii) $\dfrac{7}{8} - \dfrac{5}{6} = \dfrac{21}{24} - \dfrac{20}{24} = \dfrac{1}{24}$.

S8

$\dfrac{15}{100} \times 5.6 = 0.84$, hence answer given as £0.84.

S9

(i) £50 $\times$ 1.06 = £53;

(ii) £800 $\times$ 0.83 = £664.

S10

(i) 5.689×10^5; (ii) 5.27×10^{-4}.

E X A M   T Y P E   Q U E S T I O N S

Q1

Mr Teacher takes 24 pupils on an outing to Chester, and while there he decides to let them split into equal groups to go rowing on the river. What size boats must he be looking for?

Q2

A square playground has an area of 81 square metres. How long is each side?

Q3

Look at the following pattern and complete the next three rows.

$$1 \qquad\quad = 1 = 1^2$$
$$1 + 3 \quad\; = 4 = 2^2$$
$$1 + 3 + 5 = 9 = 3^2$$

(NEA)

Q4

1   3   8   9   10

From these numbers, write down:
(a) the prime number (*note:* 1 is NOT a prime number),
(b) a multiple of 5,
(c) two square numbers,
(d) two factors of 32,
(e) Find two numbers m and n from the list such that $m = \sqrt{n}$ and $n = \sqrt{81}$.

Q5

Last year I bought a calculator for only $\frac{2}{3}$ of its usual price of £12.45. How much did I pay for it?

Q6

When planning a garden an expert suggested that you should have $\frac{1}{10}$ of it for shrubs, $\frac{3}{5}$ of it for plants and the rest for lawn. What fraction is he suggesting ought to be lawn?

Q7

Mrs Metcalf bought a car priced at £860, and was given a reduction of 20% if she paid cash. How much did she save by paying cash?

Q8

> **LOAMSHIRE BUILDING SOCIETY**
>
> Investment Account **8.875% p.a.**
> Bonus Saver Account **8.00% p.a.**
>
> **Interest is calculated day by day and added to your account on 31 December each year.**

Loamshire Building Society advertises two savings accounts.

Miss Blake has £500 to invest for a period of 3 years and she is going to put it into one of the Loamshire Building Society accounts on 1 January. If she uses the Investment Account she will withdraw the interest at the end of each year. If she uses the Bonus Saver Account she will leave the interest to be added to capital at the end of each year.

(i)   Calculate the total interest she will receive if she uses the Investment Account.
(ii) Calculate the total interest she will receive if she uses the Bonus Saver Account.
(iii) State which scheme gives more interest to Miss Blake, and by how much.

(NEA)

Q9

(a) From the map, how much warmer is Wales than Scotland?

(b) The temperatures drop by 5 °C from those shown on the map. What is the new temperature in (i) Scotland, (ii) England?   (MEG)

Q10

The mass of an electron is 0.000 000 000 000 000 000 000 000 000 91 grams.

(a) Express this number in standard form.

(b) What would be the mass of (i) 10 electrons, (ii) 3 electrons?

Q11

Put the following information into a Venn diagram and describe each region, with the number of people it represents:

Of 16 students in a classroom, 10 enjoy Maths, 9 enjoy Chemistry, and 3 of them don't enjoy either subject.

Q12

A Road Safety Officer reported to the Head of a certain school as follows.

'I examined 100 bicycles of your pupils, looking for faults in brakes, steering and saddle heights. I found 9 cases of defective brakes, 4 of bad steering and 22 bicycles with saddles at the wrong height.'

'It is sad to reflect that although three-quarters of the bicycles had none of these faults, 4 had two faults and 3 failed on three counts. The 16 bicycles which had only saddle height faults, I have corrected straight away.'

(i) Present this information in a Venn diagram,
    using B = {bicycles with faulty brakes},
           S = {bicycles with bad steering}, and
           H = {bicycles with saddles at the wrong height}.

(ii) Evaluate $n\{B \cup S' \cup H'\}$

Interpret this result.   (NEA)

Q13

Nineteen people are employed in an office. The Venn diagram shows some details about the number who can do audio-typing (A), shorthand-typing (S), and use the word processor (W). They all have at least one of these skills.

(a) 11 people can do audio-typing. Find x.

(b) 6 people cannot do either method of typing. How many can do shorthand typing?

(c) Nobody does only shorthand-typing. Find:
    (i) how many people can use the word processor;
    (ii) how many people can both use the word processor and do shorthand-typing.

(d) Copy the Venn diagram and shade the region $W \cap A' \cap S'$. Give a brief description of this set.   (SEG)

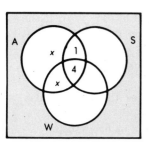

OUTLINE ANSWERS TO EXAM QUESTIONS

A1

Any factor of 24 would do apart from 1 and 24, so Mr Teacher can look for boats holding 2, 3, 4, 6, 8 or 12; that is, any of the other factors.

A2

We need the square root of 81, so do it on the calculator, putting in 81 followed by $\sqrt{}$ and you should get the answer 9. Therefore each side of the playground is 9 metres.

A3

The first sums are odd numbers, the totals are square numbers, so the next three rows will be:

$1 + 3 + 5 + 7 \qquad\quad = 16 = 4^2$
$1 + 3 + 5 + 7 + 9 \qquad = 25 = 5^2$
$1 + 3 + 5 + 7 + 9 + 11 = 36 = 6^2$

A4

(a) 3
(b) 10
(c) 1 and 9
(d) 1 and 8
(e) $m = 3, n = 9$.

A5

$\frac{1}{3}$ of £12.45 is £4.15, so $\frac{2}{3}$ will be £8.30, which is the price I paid for the calculator.

A6

We first need to add together $\frac{1}{10}$ and $\frac{3}{5}$, the $\frac{3}{5}$ needing to be rewritten as $\frac{6}{10}$. Hence $\frac{1}{10} + \frac{6}{10} = \frac{7}{10}$ of the garden would be used for shrubs and plants. Now $1 - \frac{7}{10}$ is $\frac{10}{10} - \frac{7}{10}$ which equals $\frac{3}{10}$, so the fraction left for the lawn is $\frac{3}{10}$.

A7

20% of £860 is 20 × 860p, which is 17 200p or £172, so Mrs Metcalf saved £172.

A8

(i) Since she withdraws her interest each year, she will receive the same amount of interest each year which will be $500 \times \dfrac{8.875}{100} = £44.37$ (the 0.005 will have been kept by the society). So in three years she will have received 3 × £44.37 = £133.11.

(ii) At the end of the first year she will have in her account 500 × 1.08 = £540.
At the end of the second year, she will have £540 × 1.08 = £583.20.
At the end of the third year, she will have £583.20 × 1.08 = £629.85 (the 0.006 being kept by the society).
Hence the interest gained will have been £629.85 − £500 = £129.85.

(iii) The Investment account is the greatest by £3.26.

A9

(a) The difference between 4 °C and −5 °C. Start at either point on the temperature scale and count to the other – you should get 9 °C. So, on the map, Wales would be 9 °C warmer than Scotland.

(b) To find the new temperature in Scotland, count down 5 °C from −5 °C and you come to −10 °C. To find the new temperature in England, count down 5 °C from 1 °C and you come to −4 °C.

A10

(a) There are 27 zeros, so the decimal point needs to move 28 places to the 9.1 position, making the standard form number 9.1×10^{-28}, negative since it is less than 1.

(b) (i) Multiply by 10 just by moving the decimal point one place to the right; here it would be one less zero, so standard form number will be 9.1×10^{-27}.

(ii) $9.1 \times 10^{-28} \times 3$ will be 27.3×10^{-28}, which is $2.73 \times 10 \times 10^{-28}$ which will be 2.73×10^{-27}.

A11

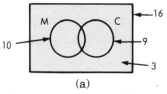
(a)

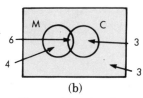
(b)

From diagram (a) we can work out that the centre will contain $(10 + 9) - 13$, which is 6.

Hence the full picture is given in diagram (b), which we can describe as illustrating

      6 students that like both Maths and Chemistry
      4 students that like Maths but not Chemistry
and 3 students that like Chemistry but not Maths.

A12

(i) You should have a Venn diagram similar to the one shown.

(ii) This is 100, it is the number of bikes that do not have only bad steering and saddles at the wrong height. (It is all but the shaded region.)

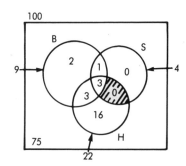

A13

(a) $2x + 5 = 11$; $2x = 6$ $x = 3$

(b) $n(A \cup S) = 19 - 6 = 13$;
    $n(S) = 13 - 6 = 7$

(c) $n(W) = 19 - 4 = 15$;
    $n(W \cap S) = 4 + 2 = 6$

(d) The diagram shows the set of people who can only use the word processor.

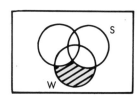

G R A D E   C H E C K L I S T

FOR A GRADE F

You should now know, and be able to describe:

> ▶ simple proportions of whole numbers up to 100, such as even, odd, square, prime, multiples and factors;

you should now understand:

> ▶ simple fractions, percentage, negative numbers;

and be able to:

> ▶ write down the square root of a square number not larger than 100;
> ▶ add and subtract decimal numbers with up to two decimal places;
> ▶ multiply and divide decimal numbers by 10 and 100;
> ▶ calculate a simple fraction of a quantity;
> ▶ add and subtract simple vulgar fractions with bottom numbers that are the same or simply related;
> ▶ convert a simple vulgar fraction to a decimal;
> ▶ express a percentage in terms of a decimal;
> ▶ calculate a given percentage of a sum of money;
> ▶ increase and reduce a sum of money by a given percentage.

FOR A GRADE C

You should also be able to:

> ▶ carry out calculations involving decimals with answers to no more than three decimal places;
> ▶ carry out calculations involving addition, subtraction and multiplication of fractions with denominators up to 10 (and 16);
> ▶ reduce a vulgar fraction to its simplest form;
> ▶ express any appropriate number in standard form;
> ▶ convert between vulgar fractions, decimals and percentages;
> ▶ express one quantity as a percentage of another;
> ▶ calculate percentage change;
> ▶ use a Venn diagram to solve a simple situation.

FOR A GRADE A

You should also be able to:

> ▶ understand set notation and interpret Venn diagrams.

STUDENT'S ANSWER - EXAMINER'S COMMENTS

QUESTION

The local cricket team has arranged to play some friendly games in Australia, combining this with a package holiday. There is a weight limit for passengers on package holidays of 20kg of luggage per person. The night before they leave the team members are asked to report to the club pavilion with their luggage for weighing. The club secretary makes the following list.

G. Boycott	20kg	B. Statham	18kg
G. Gooch	25kg	F. Trueman	19kg
D. Gower	19kg	N. Cowans	20kg
D. Randall	18kg	J. Laker	17kg
D. Bairstow	22kg	H. Larwood	14kg
I. Botham	25kg		

(i) Who should be sent home to repack?

G. GOOCH , D. BAIRSTOW , I. BOTHAM

Correct

(ii) How many kilograms overweight are each of them?

G. GOOCH WAS 5Kg OVERWEIGHT

D. BAIRSTOW WAS 2Kg OVERWEIGHT

I. BOTHAM WAS ALSO 5Kg OVERWEIGHT

(iii) The club secretary, having second thoughts, suggests that the overweight luggage could be shared among those whose luggage is not overweight. This they do and, in consequence, no player has to be sent home. Explain why?

SOME PLAYERS HAD LIGHTWEIGHT BELOW 20Kg MARK SO THEY COULD TAKE SOME OVERWEIGHT FROM THEIR COLLEGUES AND MAKE IT EVEN. THE WHOLE TEAM WAS 15 Kg LESS BUT THERE WERE 3 PLAYERS WITH EXTRA 12Kg. THIS 12Kg WAS TAKEN BY 15 Kg THEY CAN SPARE AND STILL THE TEAM HAD 3Kg TO SPARE

a good explanation

Not perfect English, but good enough to earn full marks

APPROXIMATION

G E T T I N G   S T A R T E D

We all use some kind of estimations or approximations every day. 'I'll need about £5 tonight, Dad', or 'There were about 200 at the disco tonight.' In many situations it is much more helpful to use approximations than to try to be exact. Only a Mr Spock would say, 'You've got 36 hours, 53 minutes and 8 seconds before it rains!' So in this chapter we look at the rules we use to approximate. We then apply these rules to real situations as well as to examination questions. If you fail to round off to a suitable degree of accuracy you will lose marks in the examination.

USEFUL DEFINITIONS

Metric	based on the metre as a standard of measurement
Imperial	belonging to the official British series of weights and measures

---

E S S E N T I A L   P R I N C I P L E S

1 ▷ ESTIMATION

WEIGHT

At some time we all estimate *weights*, as when we are in a supermarket buying the vegetables. A good standard to use is a bag of sugar because it weighs 1 kg or approximately 2 pounds.

WORKED EXAMPLE 1

Estimate the weight of a brick.

If you've never held a brick find one, and make a guess as to how much it weighs (try holding a bag of sugar in one hand and the brick in the other). If your guess was about 3 kilograms then you were about right.

LENGTH

Lengths of objects or distances are other common estimations. Do you know approximately how far it is to your school? Try estimating how far it is to the nearest pub or to the next town or village. (Some kind person might check your estimate on a car's mileometer.)

The units you use to estimate are important, since you would be rather foolish to try to estimate your height in kilometres or the distance travelled to school in centimetres. You have to use sensible units, like kilometres (or miles) for long distances, metres for large object lengths, e.g. cars and buses, and centimetres for small objects like pencils, books or feet.

WORKED EXAMPLE 2

Estimate the length of a bus.

A sensible unit will be metres. Try to imagine that you are walking from the back of the bus to the front. How many paces will you take (about 12?); how long is each step (about half a metre?). That makes the bus about 6 metres long.

LIQUID

How many times have you had to estimate half a pint or a litre for some 'instant' food that needs cooking? To estimate pints is quite easy, since most of us can find or recognise a pint milk bottle. But what about a litre? A litre is about $1\frac{3}{4}$ pints; an estimate of 1 litre to 2 pints is reasonable.

A millilitre is one thousandth of a litre and almost exactly the same as one centimetre cubed (1 cm^3). A good estimate for millilitres is that 1 cup holds about 200 millilitres.

2 ▷ UNITS

You need to be familiar with everyday units both metric and imperial, as well as the approximate links between them.

METRIC

You do need to know that the word 'kilo' means 'thousand' so that

1 kilogram	= 1000 grams	or	1 kg = 1000 g
1 kilometre	= 1000 metres	or	1 km = 1000 m
1 kilowatt	= 1000 watts	or	1 kW = 1000 W

Other essential units to know are:

1000 kilograms	= 1 tonne	or	1000 kg = 1 t
10 millimetres	= 1 centimetre	or	10 mm = 1 cm
100 centimetres	= 1 metre	or	100 cm = 1 m
1000 millilitres	= 1 litre	or	1000 ml = 1 l

Some mathematics syllabuses will tell you these, others will expect you to know them. It is very useful for your own use, though, that you know them all yourself.

IMPERIAL

Other common units that you should still be familiar with are:

$$12 \text{ inches} = 1 \text{ foot}$$
$$3 \text{ feet} = 1 \text{ yard}$$
$$16 \text{ ounces} = 1 \text{ pound}$$
$$8 \text{ pints} = 1 \text{ gallon}$$

You are probably used to these words and perhaps you even use these units more than the metric ones. So it is very useful to be aware of the approximate conversions. We say

2 pounds weight	is approximately	1 kilogram
3 feet	is approximately	1 metre
5 miles	is approximately	8 kilometres
1 gallon	is approximately	$4\frac{1}{2}$ litres

3 ▷ TIME

With digital watches, videos and timetables we are often faced with the problem of reading from a 24-hour clock and converting the time to our regular 12-hour clock.

A digital 24-hour clock will usually display time as 09 : 14 where 09 refers to the hour of the day and the 14 refers to the minutes after the hour. When the hour is greater than 12, take 12 off to find the afternoon hour, e.g. 15 : 25 refers to twenty-five past three in the afternoon.

SHEFFIELD – BRISTOL

Sunday

SHEFFIELD	0948
Chesterfield	1013
DERBY	1056
Burton-on-Trent	1113
BIRMINGHAM, New Street	1203
BRISTOL, Temple Meads	1457

> **This type of question will often appear on the basic and intermediate level papers**

If we wish to find how long we take to go from Sheffield to Chesterfield from the timetable then we count on from the Sheffield time of 0948 to the Chesterfield time of 1013. There will be 12 minutes up to 1000, plus 13 minutes to 1013, which gives us 25 minutes journey time.

The time from Sheffield to Derby can be found by noticing that the Derby time of 1056 has 8 more minutes than the 48 at Sheffield, with the hour just 1 more, giving a length of time 1 hour 8 minutes.

The time from Sheffield to Birmingham will be a combination of the previous two methods. Count 12 minutes up to 1000, then 2 hours to 1200, then 3 minutes to 1203, giving 2 hours 15 minutes.

WORKED EXAMPLE 3

A man who was going to catch the 0948 to Bristol as in the timetable above got up late, had a puncture and finally arrived at the railway station at 12.45 p.m. just in time to get the next train to Bristol. Estimate about what time he would get into Bristol.

From the given timetable we can work out that the train takes 5 hours 9 minutes to get to Bristol. 5 hours 9 minutes after 12.45 p.m. gives us 1754, so we could round this up to 1800, which is 6.00 p.m. early evening.

WORKED EXAMPLE 4

A train arrives in Sheffield at 2.10 p.m. after a journey lasting 6 hours 25 minutes from Tenby. At what time did it set off?

We need to work back from 2.10 p.m., or 1410. First the hours; take 6 hours off which gives us 0810, then go back 25 minutes to 0745.

EXERCISE 1

A film starts at 2150 and finishes next morning at 0115. How long does the film last?

4 > ROUNDING OFF

We often do this when estimating: we try to be as accurate as we can to start off with, and then round off the number. To round off to the nearest 10 we need to look at the unit digit; if it's below 5 we round *down*, if it's 5 or above we round *up*. For example,

21	rounds down to	20 (the 1 being less than 5)
138	rounds up to	140 (the 8 being more than 5)
285.9	rounds up to	290 (the 5 being equal to 5)
397	rounds up to	400 (the 7 being more than 5)

To round off to the nearest 100 we look at the tens digit; once again, if it's below 5 we round down, if it's 5 or higher we round up. For example,

283	rounds up to	300 (the 8 being more than 5)
5749	rounds down to	5700 (the 4 being less than 5)

WORKED EXAMPLE 5

What is the approximate weight of a milk crate containing 20 pint bottles of milk?

Each pint of milk is approximately the same weight as a 1 kg bag of sugar, so making 20 kg of milk. The plastic crate itself would also be around 1 kg in weight, but an estimate of 21 seems too specific, and so an estimated weight of 20 kg is the expected answer.

WORKED EXAMPLE 6

Maureen sorts magazines into envelopes to be posted. In a morning from 9.00 till 12.00 she manages to sort 253 magazines.
(a) Approximately (to the nearest 10) how many magazines does she sort in 1 hour?
(b) How many magazines will she sort out in a week when she works for 16 hours? (Give your answer to the nearest 100.)

(a) Use a calculator or try a division of 253 ÷ 3, and you will get an answer like 84.333. It doesn't matter what numbers we have after the unit figure of 4, since the 4 shows us that we round down to 80 (to the nearest 10). Answer then, 80 magazines an hour.
(b) We can use our answer of 80 and multiply by 16 to get 1280, which will round up to 1300 (to the nearest 100), *or* you could have used the accurate answer of 84.333 multiplied by 16, which would still round off to 1300.

5 ▷ SIGNIFICANT FIGURES

When we estimate or approximate we generally use *significant figures*.

'The crowd at the football match today was about 30 000.' This estimation is to one significant figure, since the 3 is significant and all the other digits are 0.

'His age is about 45.' This is to two significant figures, since two figures are used.

'The distance between Sheffield and Tenby is about 250 miles.' This also is two significant figures. A simple table may help you to see this more easily:

One significant figure	8	30	500	9000	0.003
Two significant figures	13	370	2100	45 000	0.071
Three significant figures	217	36.5	1.05	4710	20 800

By now you should be using a calculator competently. This means that you should be pressing the right buttons for the right calculation. But mistakes are made! So you must always be prepared to estimate your answer to see if the figure given by the calculator is about right. One way of estimating an answer to a calculation you have to do is to round every number off to one significant figure, then to do the calculation to one or two significant figures.

WORKED EXAMPLE 7

Estimate the cost of 83 packets of crisps at 17p each.

Round off 83 to 80, and 17 to 20 to give an answer of 80 × 20p = £16.00. Hence we would estimate the cost to be around £16.

WORKED EXAMPLE 8

A family went from Scotland to Cornwall covering 571 miles in $9\frac{1}{2}$ hours. What was their approximate average speed?

Round 571 to 600 and $9\frac{1}{2}$ to 10, to give an answer of 600 ÷ 10 = 60 mph.

EXERCISE 2

About how many cars can park on one side of a road that is 100 metres long?

6 ▷ DECIMAL PLACES

When we do want an accurate answer, using a calculator can often give us an answer that is *too* accurate and we need to round off the display to a suitable number of decimal places. Decimal places are the places to the right of the decimal point. A simple table may again help you to see what this means:

One decimal place	3.6	574.9	0.7	300.5	29.0
Two decimal places	137.05	2.75	31.32	219.47	8.00
Three decimal places	0.763	3.009	41.699	0.056	10.008

> errors in rounding off are one of the most common made in the exams, so be warned!

So by considering π, which when put into your calculator display is 3.141 592 7,

π to one significant figure is 3
π to one decimal place is 3.1
π to two decimal places is 3.14
π to three decimal places is 3.142.

WORKED EXAMPLE 9

Find the circumference of a circle with diameter of 8 cm. Use the formula $C = \pi D$.

Use your calculator to work out $\pi \times 8$, and you will be shown the display 25.132 741 (at least). A suitable degree of accuracy here would be to one decimal place, i.e. 25.1 cm.

WORKED EXAMPLE 10

Karen, who earned £37.60 a week, has been given a 7.8% pay increase. Calculate her new pay.

Karen's pay rise would be $\dfrac{7.8}{100} \times 37.60 = 2.9328$. As we are dealing with money this needs rounding to two decimal places, thus giving an increase of £2.93. Hence, the new pay will be £37.60 + £2.93, which is £40.53.

SUITABLE ACCURACY

When you calculate the answers to many problems, especially on the calculator, you often get an accuracy of far too many decimal places or significant figures than the problem merits. You are then expected to round off any answer to a suitable degree of accuracy. As a 'general rule' you should round off answers to no more than one significant figure extra to the figures given in the calculation.

WORKED EXAMPLE 11

A length of wood 140 cm is cut into three equal pieces. How long is each piece of wood?

140 ÷ 3 = 46.6̇6̇. But, since the largest number of significant figures in the question is two (the 140) we round off to three significant figures, giving us 46.7 cm.

GEOMETRICAL ACCURACY

Questions involving the use of trigonometry and Pythagoras, which you will meet in Chapter 11, are where this problem is most likely to be faced.

$x = 5 \sin 76 = 4.851\,478\,6$.
Since the largest number of significant figures given is two (the 76), we round off to three significant figures, giving us 4.85 cm.

By Pythagoras' rule, $x = 8.602\,325\,3$.
Since the largest number of significant figures given is one, we round off to two significant figures, giving us 8.6 cm.

7 ▷ LIMITS OF ACCURACY

If an object is measured and we are told that to the nearest centimetre it is 5 cm, then the object itself could be as small as 4.5 cm (and no smaller) or as large as 5.499 (but not 5.5 cm). Or, in other words, 4.5 cm ≤ object < 5.5 cm. Note the use of the inequality sign at each side.

WORKED EXAMPLE 12

If a marble is said to weigh 7.8 grams, what is (a) the maximum and (b) the minimum weight of 100 similar marbles?

If one marble weighs 7.8 grams, then since this is written to one decimal place we may assume that the marble can weigh between the limits 7.75 g ≤ marble weight < 7.85 g. Hence 100 of them will be within limits 775 g ≤ marble weight < 785 g. Thus,
(a) the maximum weight will be just under 785 g,
(b) the minimum weight will be 775 g.

EXERCISE 3

£5 worth of tenpenny pieces should weigh 540 grams to the nearest 10 grams. What is the minimum weight of one tenpenny piece?

SOLUTIONS TO EXERCISES

S1

Up to midnight is 2 hours 10 minutes, then 1 hour 15 minutes after midnight, giving a total length of 3 hours 25 minutes.

S2

Imagine one car: it is about 3 metres long. Allow 1 metre for space between cars and we are using 4 metres per car. Since $100 \div 4 = 25$, we would say that about 25 cars would be able to park.

S3

The least weight of the £5 will be 535 grams. Hence $535 \div 50$ gives the minimum weight of 10.7 grams.

EXAM TYPE QUESTIONS

The majority of this topic is tested within other topics so that it is put into its proper context.

Q1

The picture shows a woman of average height standing next to a lamp post.
(i) Estimate the height of the lamp post.
(ii) Explain how you got your answer.   (NEA)

Q2

On a foreign holiday a motorist was warned that the speed limit was 130 kilometres/hour. What is this speed to the nearest 10 miles per hour?

Q3

One afternoon from 1 o'clock to 5 o'clock a sausage machine produced 3551 sausages. Each sausage weighed around 60 grams.
(a) How many sausages would you say the sausage machine produced each hour?
(b) What approximate weight of sausage meat would you expect to have to use during this particular afternoon?

Q4

Given that $y = \dfrac{9}{x}$, complete the following table of values, stating the values, where appropriate, to two decimal places.

x	1	2	3	4	5	6	7	8	9
y	9		3				1.29	1.13	

(NEA)

Q5

Rick set off at 11.25 a.m. on his motorbike up the M1. He averaged about 50 miles per hour. At approximately what time would you expect him to arrive at his destination 120 miles away?

Q6

(a) Measure the length of this line: ─────────────
(b) What will be the area of a square with sides this long?
(c) What is the maximum percentage error you could have in (a) but still be correct?

Q7

A motorist buys 26 litres of petrol at the garage which displays this sign. How much will he be asked to pay for the petrol?   (NEA)

★★★★   *per litre*
43.4p
Esso

OUTLINE ANSWERS TO EXAM QUESTIONS

A1

(i) 15 feet (or 4.5 metres).
(ii) The height of the woman was estimated at 5 feet or 150 cm, then the lamp post was estimated to be 3 times higher.

A2

Since 8 kilometres is approximately 5 miles, 130 kilometres will be $\left(\dfrac{130}{8} \times 5\right)$ miles, which is 81.25. Therefore, to the nearest 10 this will be 80 mph.

A3

(a) 3551 ÷ 4 hours = 887.75, so we would say the machine made about 900 sausages an hour (890 would be too precise an approximation).
(b) The most accurate we can be is 3551×60 grams = 213 060 grams, which is 213.06 kilograms, which we can round off to 200 kilograms.

A4

x	2		4	5	6		9
y	4.5	...	2.25	1.8	1.5	...	1

A5

120 miles at 50 mph will give us 120 ÷ 50 = 2.4 hours, which is about $2\frac{1}{2}$ hours. Rick's expected time of arrival then is 11.25 a.m. $+ 2\frac{1}{2}$ hours, which is about 2.00 p.m. We might say 'we expect him just before two'.

A6

(a) The line can only really be measured accurately to the nearest millimetre which will be 3.8 cm (or 38 mm).
(b) The area will be $(3.8)^2 = 14.44$, but to three significant figures this is 14.4 cm².
(c) The greatest 'error' is to be 0.05 out (i.e. 3.75 cm or 3.85 cm). So, $\dfrac{0.05}{3.8} \times 100 = 1.3\%$ error.

A7

$26 \times 43.4p = 1128.4p = £11.28$.

G R A D E   C H E C K L I S T

FOR A GRADE F
You should know:

what the phrase 'rounding off' means;

the relationships between common metric units and their approximations in imperial units;

the 24-hour clock system;

you should understand:

why we round off when approximating;

a timetable written in either 12- or 24-hour clock system;

and be able to:

estimate answers by making use of simple approximations;

convert between 12-hour and 24-hour systems;

work out the interval between two times given in the same system;

work out the finishing time given the starting time and the duration of a journey or task;

work out the starting time of a task given the finishing time and the duration of the task;

read a timetable;

use a timetable to plan a journey;

estimate lengths, weights and capacities of objects encountered in the home;

use a limited range of imperial units in contexts in which they are still commonly used.

FOR A GRADE C
You should also know:

what significant figures and decimal places are

and be able to:

use a calculator competently;

estimate the approximate value of an arithmetical expression;

round the result given by a calculator to an appropriate number of decimal places;

write whole numbers to a required degree of accuracy.

FOR A GRADE A
You should understand:

the limits of accuracy, and appreciate how much 'out' a given figure can be

and be able to:

give an answer to a problem to a suitable degree of accuracy.

STUDENT'S ANSWER – EXAMINER'S COMMENTS

QUESTION

Kleenup washing powder is sold in 800g packets which cost 87p each. The instructions on the packet state:

1 cup of powder weighs approx. 100g (3½ oz)

Quantity to use:

		Hard Water	Soft Water
Top loading automatics			
8 gallon size		3 cups	2 cups
10 gallon size		4 cups	3 cups
15 gallon size		5 cups	4 cups
Soak and handwash			
Average sink (5 gallons)		2 cups	1½ cups
Average bowl or bucket (2 gallons)		¾ cup	½ cup

(i) What is the approximate cost of the washing powder for an 8 gallon size wash in a soft water area?

correct method shown, but no final answer →

8 GALLON SIZE = 2 CUPS (S.W. AREA) = 200g   800g = 87p

200g = 87/4

Joseph lives in a hard water area and always uses the soak and handwash method, using the sink every other week for a large wash and a bowl twice every week for a small wash. On average how many weeks will one packet of kleenup washing powder last Joseph?

TWICE A WK. = ¾ CUP x 2 = 1½ CUP = 150g

ONCE EVERY OTHER WEEK = 2 CUPS = 200g

a correct answer, but the method does not clearly show *why* the answer is 3 weeks. You will therefore lose marks even though you have the right answer. →

WK 1 = 350g , WK 2 = 150g , 3 WK = 350

∴ ONE PACKET OF KLEEN UP WILL LAST

3 WEEKS .

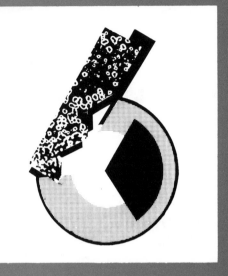

APPLICATION

It is an important part of the GCSE examining system that as many questions as possible have some 'real-life' application. Throughout this book, many of the settings used are those of real-life situations. However in this chapter we focus on the *applications* which are specially highlighted in all the Examination Board syllabuses.

To answer many of the examination questions you must be able to obtain information from a *table* or *chart*, so we illustrate ways of doing this. We also include the different types of *scales* with which you should be familiar. A lot of this chapter uses ideas of number and approximation that have already been met in previous chapters.

USEFUL DEFINITIONS

Deposit	money paid as an initial payment on hire purchase agreement
Discount	a deduction from the usual price
Interest	a charge for borrowed money
Premium	the amount paid for an insurance contract

HIRE PURCHASE

DISCOUNT

INTEREST

TYPES OF INTEREST

PROFIT AND LOSS

TAX

WAGES AND SALARIES

USE OF TABLES AND CHARTS

READING DIALS AND SCALES

E S S E N T I A L   P R I N C I P L E S

1 > HIRE PURCHASE

Many people buy on what is called the 'never never', or 'buy now pay later'. These are alternative phrases for hire purchase (HP), which is a convenient way to spread out a large payment over a period of time. It usually requires a *deposit* to be paid before the goods are taken out of the shop, and a promise (contract) to pay so much a week or a month for a number of weeks or months.

WORKED EXAMPLE 1

TV cash price £450
HP 10% deposit and 18 monthly payments of £24
Calculate how much higher the HP price is than the cash price.

The deposit is 10% of £450, which is $\dfrac{10}{100} \times 450 =$ £45

The total of the 18 payments is $18 \times £24 = £432$
So the total HP price is $£45 + £432 = £477$
Therefore the HP price is higher by $£477 - £450 = £27$

EXERCISE 1

Calculate the total HP price on a microwave bought for £35 deposit and 9 monthly payments of £24.

2 > DISCOUNT

Discount is an amount of money that the shopkeeper will give you for buying goods in a particular way. The discount is not paid in actual cash but is a way of reducing the price of the goods to you. Often people are given a discount if they pay cash for their purchase, or if they open an account, or even if they work for a particular firm or are members of a particular union.

WORKED EXAMPLE 2

James is a 'young owl', and as such is entitled to a 5% discount on any purchase at the 'Owls souvenir shop'. How much would he pay for a badge priced at £2.60?

The 5% discount on £2.60 will be $\dfrac{5}{100} \times 2.60 = £0.13$,

so James will pay $£2.60 - £0.13 = £2.47$.

EXERCISE 2

What is the price paid for a £310 washer with a discount of 10%?

3 > INTEREST

Interest is what we call the amount of money someone will give you for letting them borrow your money, or what you pay for borrowing money. So you can be given interest and you can also be asked to pay interest.

Banks and building societies give you interest if you let them borrow some of your money. For example, a well-known bank will pay you 6% interest per annum (that is per year), so if you leave £20 in their bank for one year they will pay you $\dfrac{6}{100} \times £20 = £1.20$.

WORKED
EXAMPLE 3

I. Payupp Limited offer loans of £1000, with repayments of £68.75 each month for 16 months. What is the interest paid on the loan?

16 payments of £68.75 = £1100, which is £100 more than the amount borrowed. Therefore, £100 is the interest paid on the loan.

4 ▷ TYPES OF INTEREST

You have just met the idea of interest, but there are two types of interest. One is *simple* interest and the other is *compound* interest. It will be 'interesting' to see the difference between the two!

SIMPLE INTEREST

Simple interest is calculated on the basis of having a principal amount, say, P, in the bank, for a number of years T, with a rate of interest R. There is then a simple formula to work out the amount of interest your money will earn. It is $\text{SI} = \dfrac{PRT}{100}$, which means that simple interest is found by multiplying the principal by the rate, then by the time, and then dividing by 100.

WORKED
EXAMPLE 4

John had £16.40 in an account that paid simple interest at a rate of 9%. Calculate how much simple interest would be paid to John if he kept the money in the account for 5 years.

The principal is £16.40, the rate is 9% and the time is 5 years. Hence, using the formula, $\text{SI} = \dfrac{16.40 \times 9 \times 5}{100} = £7.38$.

EXERCISE 3

What is the simple interest on £250 over 3 years at 8.75% interest rate?

COMPOUND INTEREST

Compound interest is the interest most likely to be paid to you by banks and building societies. It is based on the idea of giving you the simple interest after 1 year and adding this on to your principal amount (sometimes every 6 months). The money then 'grows' more quickly than it would with just simple interest.

WORKED
EXAMPLE 5

Elsie won £60 in a beauty contest and put it into a building society account that paid 8% compound interest annually. How much would she have in this account if the money was left there for 3 years?

After 1 year the simple interest will be $\dfrac{60 \times 8 \times 1}{100} = £4.80$, which is added to the account. So the second year starts with a principal amount of £64.80. At the end of the second year the simple interest will be $\dfrac{64.80 \times 8 \times 1}{100} = £5.184$, which would be rounded off to £5.18 and added to the £64.80 to give a new principal amount of £69.98. At the end of the third year the simple interest will be $\dfrac{69.98 \times 8 \times 1}{100} = £5.5984$, which would be rounded off to £5.60 which, when added to £69.98, will give a final figure of £75.58.

EXERCISE 4

£60 is invested for 3 years at a rate of 9% compound interest. How much interest is this altogether?

5 ▷ PROFIT AND LOSS

A shopkeeper usually sells his goods for more than he paid for them and this is called his *profit*. If he sells for less than he paid then he makes a *loss*.

WORKED EXAMPLE 6

> The examiner will expect you to include the 25% profit margin you have already been told about.

Jim Karna sold second-hand cars and always tried to make about 25% profit on the cars he bought for resale. He bought a nice yellow Hillman Hunter for £650. What price would you have expected him to sell it for?

25% of £650 is £162.50, which added to £650 gives £812.50. We would expect this price to be rounded down to £800 as the selling price

WORKED EXAMPLE 7

The Hot House Fuel Company made a total loss of 8% on their transactions of £500 000 in 1986. How much loss did they make?

The loss is just 8% of £500 000, which is £40 000.

6 > TAX

Tax is the amount of money that a government tells its people to pay in order to raise sufficient funds for that government to run the country. The tax system is generally complicated, but you are only expected to be familiar with two major types of tax which we will look at here.

VAT

VAT (or value added tax) is the tax put on to the price of goods sold in shops, restaurants, etc., then paid to the government. The tax is usually a *percentage* and can vary from year to year. The tax also depends on the *type* of goods sold, with some goods, such as books, having a zero rate of VAT, and others the usual 15% rate.

WORKED EXAMPLE 8

In 1988 an electrical shop bought cassette players from a warehouse in boxes of 10 for £90. To work out their selling price the shopkeeper added on 37% for his profit, then added on the VAT. The VAT on these goods in 1988 was 15%. Calculate the selling price and the amount of VAT on each cassette player.

The shopkeeper bought each player for £90 ÷ 10 = £9.
His profit of 37% gives him a profit of £3.33 on each one.
The VAT on this new total of £12.33 at 15% is £1.8495 which, when rounded off, becomes £1.85. The selling price is now £14.18, including VAT of £1.85.

INCOME TAX

Income tax is the type of tax that everyone who receives money for working or from investments has to pay to the government. Here again the amount can change every time the government decides to alter it. To calculate how much tax you should pay you first need to know the rate of tax (a percentage) and your personal allowances.

Personal allowances are the amounts of money you may earn before you start to pay any tax; they are different for single men and married men, and for women in different situations, and can be increased for quite a number of different reasons.

You only pay tax on your *taxable pay*, which is found by subtracting your personal allowances from your actual annual pay. If your personal allowances are greater than your actual pay then you would pay no income tax.

The *rate of tax* is expressed either as a percentage, for example 30%, which means that you would pay a tax of 30% on your taxable income, or it may be expressed at a certain rate in the £. For instance, if it was 27p in the £, you would pay 27p for every £1 of taxable pay (which is equivalent to 27%).

When the rate of tax is 27%, find the income tax paid by Mr Dunn who earns £18 600 per annum and has personal allowances of £5800.

Mr Dunn's taxable pay is   £18 600 − £5800 = £12 800.

The rate of tax is 27%, so he would pay $\frac{27}{100} \times £12\,800 = £3456$.

EXERCISE 5

Mr Palfreyman is paid a salary of £14 492 per annum which he receives in 12 monthly payments. He has personal allowances amounting to £3860. If tax is payable at the rate of 27p in the £, calculate his monthly pay.

7 ▷ WAGES AND SALARIES

WAGES

Wages are the amounts people earn in a week for working. Wages usually vary with the number of hours worked. People normally have a basic week – that is, a set number of hours to be worked in a week – and receive a basic pay calculated on an hourly basis. For example, Jane is paid £3.20 per hour for a basic week of 38 hours, so her normal wage for the week is £3.20 × 38 = £121.60.

Any extra time worked is called overtime, and is paid by various overtime rates:

time and a quarter   is basic hourly rate × $1\frac{1}{4}$
time and a half   is basic hourly rate × $1\frac{1}{2}$
double time   is basic hourly rate × 2.

Ethel, whose basic week consists of 32 hours at £2.10 per hour, works for 40 hours one week. The overtime rate is time and a half. Calculate Ethel's wage for that week.

The basic week is 32 × £2.10, which is £67.20. The overtime of 8 hours (40 − 32) will be 8 × 1.5 × £2.10, which is £25.20. Therefore £67.20 + £25.20 will give you Ethel's wage for the week, which is £92.40.

SALARIES

Salaries are the amounts people earn in a year. They are usually paid in either 12 monthly payments throughout the year, or 13 payments made every 4 weeks. So when people say they are paid monthly you need to know if they are paid for 12 'calendar' months, e.g. paid on 5 January, 5 February and so on every month, or if they are paid every 4 weeks (a lunar month) which will give them 13 regular payments throughout the year.

Joe has a salary of £16 380. What would be the difference in the amounts received if Joe was paid (a) each calendar month, or (b) every 4 weeks?

(a) The pay for each calendar month would be £16 380 ÷ 12 = £1365.
(b) The pay every 4 weeks would be £16 380 ÷ 13 = £1260. The difference therefore would be £105 per payment.

PIECEWORK

Piecework is what we call payment for each piece of work done. This means that some people are paid purely for the amount of work they actually do! For example, Ned is paid 68p for every complete box he packs. If in a week he packs 160 boxes he will be paid 160 × £0.68 which is £108.80.

8 > USE OF TABLES AND CHARTS

We are confronted all the time with many types of tables and charts, from bus timetables to post office charges, and they are easy to read if you look at them in a clear logical way. You used a bus *timetable* in the previous chapter; the extract shown here is a *tidetable*.

This *tidetable* shows the approximate times of the high tides for the Sundays in 1988. You will read that the high tides on 5 September will be at 0354 which is 6 minutes to 4 in the morning, and at 1605 which is 5 minutes past 4 in the afternoon.

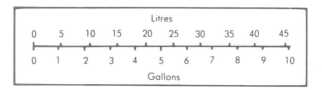

This is a different kind of table. It is used in garages to help you *convert* litres to gallons, and the other way round. You can read off from the chart that 30 litres is approximately 6.6 gallons, or that 8 gallons is approximately 36 litres.

1 Letter Post

Rates for letters within the UK and from the UK to
the Isle of Man, the Channel Islands and the Irish Republic

Weight not over	First Class	Second Class	Weight not over	First Class	Second Class
60 g	17 p	12 p	400 g	69 p	52 p
100 g	24 p	18 p	450 g	78 p	59 p
150 g	31 p	22 p	500 g	87 p	66 p
200 g	38 p	28 p	750 g	£1.28	96 p
250 g	45 p	34 p	1000 g	£1.70	Not admissible
300 g	53 p	40 p	Each extra 250 g or part thereof	42 p	over 750 g
350 g	61 p	46 p			

To post a letter weighing 275 g, you need to look at the line 'not over 300', since the previous line is lower than the 275. This letter then will cost you 53p for first class post or 40p for second class.

Monthly insurance premiums				
Age (next)	£1000	£2000	£5000	£10 000
20	1-60	3-15	7-78	15-50
25	1-85	3-40	8-05	15-75
30	2-35	3-95	8-75	16-60
35	3-40	5-10	9-95	17-90
40	6-50	9-80	18-54	33-70
45	15-70	20-56		

For female, subtract 5 years from current age.

You can see from this *insurance table* that the costs vary for different ages and amounts of insurance required. For instance, you will see that it costs £9.95 per month to insure a man aged 33 for £5000. We call this monthly amount the *premium*.

EXERCISE 6

What is the monthly premium for a woman aged 40 to be insured for £2000?

DIALS

We sincerely hope that you can read a clock (well, it's time you could)! But many meters use 'anticlockwise dials'. Look at this one, which is from an electricity meter.

Although the faces are different the hand on each clock goes from 0 to 1 to 2 and so on up to 9, then to 0 again. Hence the correct reading of this meter will be 21 397. Take a close look at the 1000 and the 10 dials and confirm the reading.

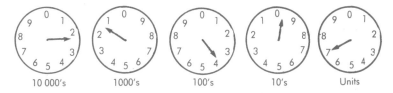

10 000's   1000's   100's   10's   Units

EXERCISE 7

See if you have an electric meter like this, and if so then read it.

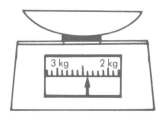

SCALES

This is a set of *weighing scales*; notice how the scale reads from right to left. Between each kilogram the space is divided into ten parts, and as one tenth of 1 kg is 100 grams, each small line represents 100 g. You can see that the pointer is on the fourth line between the 2 kg and 3 kg marks so the object we are weighing is 2 kg 400 g, or 2.4 kg.

This is the sort of *scale* you might find on a *map*. It indicates that 2 cm on the map represents 1 km in reality. The space between each kilometre is divided into ten equal parts, so each small line will represent one tenth of a kilometre which is 100 metres. Therefore a distance on the map of 3 cm will represent 1.5 km, and a distance of 3.6 cm on the map will represent 1.8 km.

Notice how on these weighing scales the space between each kilogram is divided into five parts, each line representing one-fifth of a kilogram which is 200 grams. Hence the pointer is pointing to 3 kg 600 g or 3.6 kg.

Now look at these weighing scales. The space between the kilogram is divided into ten large parts (longer lines), each one now representing 100 g (or 0.10 kg), and each of these spaces is divided into two parts, each one 50 g (0.05 kg). The pointer on the diagram is pointing to 0.25 kg.

EXERCISE 8

Have a look at the scales in your kitchen, and look in the shops to see how many different types there are in use today.

This part of the syllabus could also be assessed in a practical examination. You will need to find out whether your syllabus contains such a practical assessment or not.

SOLUTIONS TO EXERCISES

S1

£35 + (9 × £24) = £251.

S2

$£310 \times \dfrac{10}{100} = £31$, this is the discount, hence the price paid will be £310 − £31 = £279. (A quicker, most acceptable way, would have been to evaluate £310 × 0.9, which also gives £279.)

S3

$SI = \dfrac{250 \times 8.75 \times 3}{100} = £65.62$ (rounding off in this case will always be down).

S4

After 1 year the total in the account is   £60 × 1.09 = £65.40
After 2nd year the total in the account is   £65.40 × 1.09 = £71.28
After 3rd year the total will be   £71.28 × 1.09 = £77.69
Hence the interest is   £77.69 − £60 = £17.69

Note, some societies would calculate the final figure to be 60 × (1.09)³ = £77.70.

S5

Mr Palfreyman's taxable pay will be £14 492 − £3860, which is £10 632. Tax paid over the year will be 27p × £10 632, which is £2870.64, leaving a net salary over the year of £11 621.36. This divided by the 12 calendar months will make his monthly pay £968.44.

S6

£5.10.

EXAM TYPE QUESTIONS

Q1

How much will it cost for the piece of cheddar cheese on the scales if the cheese is 68p per kg?

Q2

The dials on the electricity meter at the home of Mrs Sing on 31 December are as shown. What reading is shown by the dials?

1000   100   10   1

Q3

Tina, who was paid £4.80 an hour, needed to earn an extra £50 one week for repairs to her car. She was able to work overtime at the rate of time and a quarter. How many whole hours of overtime will she need to work in order to earn this £50?

Q4

A lorry driver travels from Birmingham to Bristol and then on to Brighton at an average speed of 45 mph.
How long does the journey take? (Give your answer to the nearest hour.)

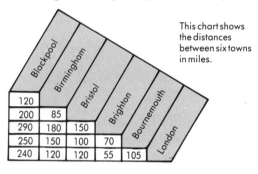

This chart shows the distances between six towns in miles.

	Blackpool	Birmingham	Bristol	Brighton	Bournemouth	London
120						
200	85					
290	180	150				
250	150	100	70			
240	120	120	55	105		

Q5

In 1985 Income tax, at the rate of 30%, was deducted from an examiner's fee of £30 for setting a question paper. How much did the examiner receive?

Q6

Travel Protector Insurance issued the following table of premiums for holiday insurance in 1985.

Find the premium paid by Mr Jones, holidaying in Blackpool with his wife and three children, aged 2, 9 and 15, from 2 August to 16 August.   (NEA)

> Read the question and the table carefully. Then answer the question set and *not* your own.

Winter Sports
Cover is available at 3 times these premium rates.

(Source: *Travel Protector Insurance*, published by National Westminster Bank PLC.)

Fig. 6Q.4
Premiums
per injured person

Period of Travel	Area 'A' UK†	Area 'B' Europe	Area 'C' Worldwide
1–4 days	£3.60	£5.40	£16.90
5–8 days	£4.50	£7.80	£16.90
9–17 days	£5.40	£9.95	£21.45
18–23 days	£6.30	£12.30	£27.95
24–31 days	£7.20	£15.25	£32.50
32–62 days	—	£24.10	£44.20
63–90 days	—	£34.50	£53.95

†Excluding Channel Islands
Discount for Children
Under 14 years at date of Application—
20% reduction
Under 3 years at date of Application—
Free of charge

Q7

The following information is from the 1984–85 Rates of Tax leaflet:

Basic rate	30% for taxable incomes between	£1	and £15 400
Higher rates	40% for taxable incomes between	£15 401	and £18 200
	60% for taxable incomes between	£18 201	and £23 100

In other words, in that year you would have paid tax at a rate of 30% on your first £15 400 of taxable income and at a rate of 40% for the next £2800, and 60% for the next £4900. Sir Keith Jarvis was overheard one evening to say 'I paid £6820 tax during the year 1984–85'. What was his taxable income?

OUTLINE ANSWERS TO EXAM QUESTIONS

A1

The scale reading is 2.7 kg, so the cost of the cheese is 2.7 × 68p = £1.836, which would be rounded off to £1.84.

A2

7901.

A3

Tina's hourly rate, with time and a quarter included, will be £4.80 × 1.25 = £6.00. The number of hours overtime she needs to work at this rate will be 50 ÷ 6 = 8.33˙, and this, rounded up to the nearest hour, will be 9.

A4

Distance travelled will be 85 + 150 = 235.
Time = distance ÷ speed = 235 ÷ 45 = 5.22
$$= 5 \text{ hours to nearest hour.}$$

A5

30% of £30 = $\dfrac{30}{100}$ × £30 = £9. Hence he received £30 − £9 = £21. Or a quicker way, he received 70% (100 − 30) of £30 = 0.7 × £30 = £21.

A6

The holiday is in UK, Area A, premium is £5.40.
Premiums are: Mr (£5.40); Mrs (£5.40); child aged 15 (£5.40).
Child of 9 = £5.40 × $\dfrac{80}{100}$ (20% reduction) = £4.32.
Child of 2 free.
Hence total premium = (3 × £5.40) + £4.32 = £20.52.

A7

The maximum tax at the basic rate was 30% of £15 400 which is £4620, so the taxable income was higher. The maximum tax at the rate of 40% was 40% of £2800 which is £1120, still not enough to top Sir Keith's £6820, so the remainder of the tax was paid at the rate of 60%. This remainder will be £6820 − £(4620 + 1120) which is £1080, representing 60% of the income higher than £18 200. As 60% represents £1080, then 1% represents £1080 ÷ 60 = £18. Hence 100% will be £1800. The total taxable income, therefore, would be £18 200 + £1800 which is £20 000.

G R A D E   C H E C K L I S T

FOR A GRADE F

You should now know:

what meters are, and what they tell you;

what is meant by hire purchase, deposit, interest, tax, discount, profit and loss;

you should understand:

what is meant by wage, salary and piecework;

and be able to

read accurately meters, clocks and dials;

calculate total costs of loans, hire purchase and insurance;

calculate wages or salary from given information;

calculate amounts of tax payable by different people.

FOR A GRADE C

You should also understand:

the differences between simple and compound interest

and be able to:

read accurately meters and scales with any size markings;

work out step-by-step compound interest over a few stages.

FOR A GRADE A

You will need to be able to cope with the more involved situations as in Q7.

STUDENT'S ANSWER - EXAMINER'S COMMENTS

QUESTION

For a motorway lamp, the length of the motorway lit by the lamp is equal to two thirds of the height of the lamp

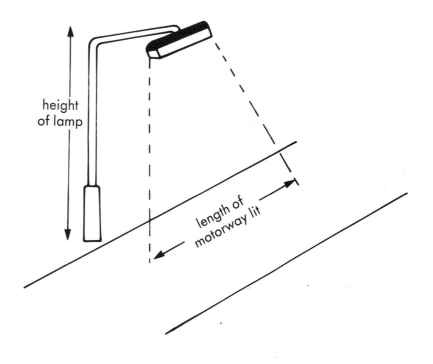

(a) Copy and complete the following table.

LAMP	HEIGHT (metres)	LENGTH LIT (metres)
X	6	4
Y	9	. 6

Good, all correct

(b) Two X lamps are positioned 6 metres apart. What length of motorway is unlit between the lamps?

Poor answer as it is wrong and there is no indication of *how* the answer has been found.

> 1   METRE

(c) Two Y lamps are positioned 5 metres apart. Draw a diagram showing clearly the section of motorway lit by *both* lamps. How long is this section?

again, the answer is incomplete. There are no measurements.

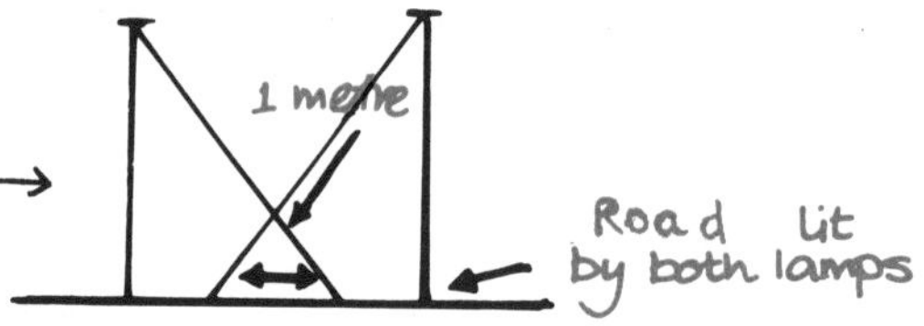

1 metre

Road lit
by both lamps

(d) The cost of a lamp (in pounds) is given by

Cost = 700 + (175 × Height) where the height is measured in metres.
Find the cost of lamps X and Y

good, both parts are correct

$$X = 700 + (175 \times 6) = £1,750$$
$$Y = 700 + (175 \times 9) = £2275$$

(e) A planning officer is trying to decide what lamps to buy for a new motorway junction of total length 100 metres. She has to make sure that the whole length of the junction will be lit and that the total cost is as low as possible.
Should she buy lamps X or lamps Y. Explain your answer fully.

$$100\,m. \quad COST \ OF \ X = \frac{100}{4} = 25 \times 1750 = £43,750$$

Good answer, except that the 16.67 should have been rounded up in the calculation.

$$100\,m. \quad COST \ OF \ Y = \frac{100}{6} = 16.67 \times 2275$$
$$= £37916.5^-$$

IT WILL BE CHEAPER TO BUY Y LAMPS

G E T T I N G   S T A R T E D

A lot of everyday arithmetic is to do with *ratio*, though we tend to deal with it in a common-sense sort of way. Ratio is a comparison between two amounts, often written with 'to' or a colon (:). One part milk *to* two parts water is a ratio. Many mathematical problems are sorted out by the formal use of ratio, which has strong links with fractions, and ultimately with algebra.

USEFUL DEFINITIONS

Enlargement	where all the respective dimensions of two shapes are in the same ratio, then each shape is an enlargement of the other
Similar	two shapes are similar if one is a mathematical enlargement of the other
Scale factor	the ratio which links two similar figures
Proportion	the relation of one number with another to form a ratio
Rate	a fixed ratio between two things
Cube root	a number which when multiplied by itself twice is a given number, the symbol of which is $\sqrt[3]{}$. Example: the cube root of 8 is 2 since $2 \times 2 \times 2 = 8$.

E S S E N T I A L   P R I N C I P L E S

1 ▷ RATIO

You mix things in certain *ratios* every day. Tea is often made with one part of milk mixed with about eight parts of tea from the pot. Of course this mix varies with taste, but certain mixtures need to be quite accurate. Take, for example, dried baby food. The directions on a packet I used read 1 teaspoon of dried food to 2 teaspoons of boiled water. This mix made 3 teaspoonsful of nice food for the baby! As the baby grew he wanted more, and I would use a bigger spoon while my wife would double up the ratio given. The initial ratio of 1 to 2 was now increased to 2 to 4 which, although being more in total quantity, was in the same ratio as 3 to 6 would be to make an even greater quantity.

WORKED EXAMPLE 1

A well-known recipe for pancake mix is: 1 egg, 6 dessertspoons of flour, $\frac{1}{2}$ pint of milk and a pinch of salt. This will make enough mix for 4 pancakes.
(a) What will be the recipe for 8 pancakes?
(b) What will be the recipe for 6 pancakes?

(a) As the recipe given is for 4 pancakes, then just double the quantities for 8. So the recipe will be: 2 eggs, 12 dessertspoons of flour, 1 pint of milk and 2 pinches of salt.
(b) This time you need to add half as much again to the original recipe. This presents problems with the egg, and although $1\frac{1}{2}$ eggs is what you ought to use, I would use 2 eggs, so the recipe will now be: 2 eggs, 9 dessertspoons of flour, $\frac{3}{4}$ pint of milk and $1\frac{1}{2}$ pinches of salt.

EXERCISE 1

Here is the recipe for a cake: 3 eggs, 200 grams of flour, 100 grams of margarine, 150 grams of sugar and a pinch of salt. This will make enough for 12 people. Mavis wants to make this cake but only has two eggs, so she makes the largest cake she can to the recipe given.
(a) For how many people will this smaller cake be enough?
(b) Write down the recipe for this smaller cake.

WORKED EXAMPLE 2

At a camp a guide leader bought a 1 litre bottle of concentrated orange juice. On the bottle it gave directions to make up with water in the ratio of 1 part concentrated orange to 6 parts water. How much orange juice will be made up altogether using all the concentrate?

7 litres of orange will be made altogether 1 litre of the concentrate plus 6 litres of water.

OTHER USES

This idea of ratio is also met with on a building site, where concrete mix is made using sand and cement in the ratio of 4 to 1. (You still then have to mix in the water to give you the ready-to-use concrete.)

WORKED EXAMPLE 3

You are told to get 100 kg of concrete mix ready with sand and cement in the ratio of 4 to 1. How much of each part, sand and cement, will you need?

By adding the 1 to the 4 you find you need 5 parts for the ratio. 100 kg ÷ 5 parts = 20 kg, so 1 part is 20 kg. Hence 4 parts of sand to 1 part of cement will be 80 kg sand to 20 kg cement.

EXERCISE 2

In 1985 Anna and Beth invested in a new company to make special baby seats for cars. Anna invested £900 and Beth £500. They decided that the end-of-year profits would be divided between them in the ratio of their investment. At the end of 1985 they had made a profit of £12 000. How much of the profit would each woman receive?

MAP SCALES

Maps very often have the scale written down as a ratio, for example 1 : 50 000, which means that 1 cm on the map would represent 50 000 cm in reality (which is 500 metres or 0.5 kilometre).

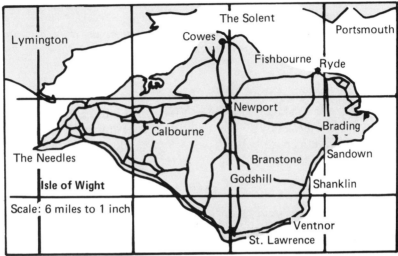

Fig. 7.1

| **WORKED EXAMPLE 4** | The scale on Fig. 7.1 is approximately in the ratio of 1 : 400 000. Estimate the actual distance between Cowes and Newport to the nearest kilometre. |

Using a ruler to measure round the slight curve, the distance on the map is 2 cm, hence the actual distance is $2 \times 400\,000 = 800\,000$ cm. Divide by 100 to change to 8000 metres, then divide by 1000 to change to kilometres, giving 8 km.

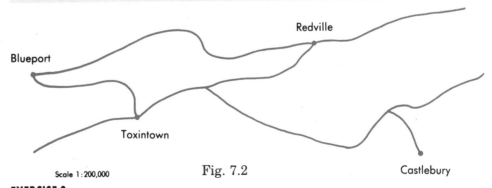

Fig. 7.2

EXERCISE 3

Use the map in Fig 7.2 with the given scale to estimate the distance of the shortest journey from Blueport to Castlebury.

2 ▷ **SCALE FACTOR**
If you enlarge a photograph then you will find that both the length and the breadth have been increased by the same *scale factor*. That is, a number by which the original lengths are multiplied to find the enlarged lengths.

The scale factor of this enlargement is 2, since the original lengths have been multiplied by 2 to find the enlarged lengths.

Model villages and model trains also have scale factors. Many model railways use the scale factor $\frac{1}{100}$ or 0.01, so reducing the size from the original but still *multiplying* the original length by the scale factor.

WORKED EXAMPLE 5

Mr Shuttleworth built a model of his church using a scale factor of $\frac{1}{20}$ or 0.05.
(a) How big is the model front door if the actual front door is 2 m high?
(b) How high is the actual roof if the model roof is 40 cm high?

(a) The model door will be the original height multiplied by $\frac{1}{20}$ (or 0.05). Change the 2 m to 200 cm, because the answer will obviously be in centimetres. So the model door will be $200 \times \frac{1}{20}$ or $200 \div 20$ or 200×0.05– each way will give the answer of 10 cm high.
(b) We need to use the scale factor the other way round this time, since we are going from the model height to the actual height. So the actual height of the roof will be 40 cm $\times$ 20 or 40 $\div$ 0.05–both ways will give the answer of 800 cm, which is 8 metres.

The scale factor may often be given as it is on maps, that is, as a ratio. For instance, in this example the scale factor could have been given as 1 : 20.

3   BEST BUYS

When shopping we are often faced with deciding which jar of the same product is the best buy. We usually do this either by finding the cost per unit weight or the weight per penny.

WORKED EXAMPLE 6

Which of the following tins of beans would represent the best buy?

It is perhaps easier to divide the bigger numbers by the smaller ones, so we will divide the weight by the cost to tell us what weight we get per penny.

Tin A: $250 \div 23 = 10.869\ 565$ grams per penny.
Tin B: $275 \div 26 = 10.576\ 923$ grams per penny.
Tin C: $230 \div 22 = 10.454\ 545$ grams per penny.

So we see that tin A gives most weight per penny and hence represents the best buy.

WORKED EXAMPLE 7

Which of the following represents the best buy of butter?

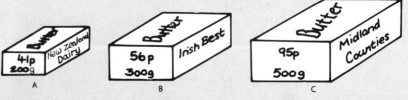

It is convenient here to calculate the cost per 100 grams.

Packet A will be 41p $\div$ 2 = 20.5p per 100 grams.
Packet B will be 56p $\div$ 3 = 18.666p per 100 grams.
Packet C will be 95p $\div$ 5 = 19p per 100 grams.

You could very well be asked to give a reason like this to justify your answer

So we see that packet B cost less per 100 grams, hence representing the best buy.

EXERCISE 4

Tom used the 1500 ml bottle of 'Weedo' on his lawn which had an area of approximately 12 square metres. He suggested to Ken, who had a lawn of approximately 20 square metres, that he also should use 'Weedo'. How much should Ken buy and what will be the best way for him to buy it?

4 ▷ PROPORTION

There are two types of *proportion* that we will consider here: *direct* and *inverse*.

DIRECT PROPORTION

Direct proportion is when there is a simple multiplying connection between two things. For example, if one marble weighs 20 g, then the weight of any number of marbles is found by multiplying the number of marbles by 20 g. This is because the weight of the marbles is in direct proportion to their number.

We use this idea if we know that the cost of, say, 5 coke drinks is £1.40, and we wish to find the cost of 12 coke drinks. We need to find the multiplying connection first, which will be the cost of 1 coke drink, and if 5 cost £1.40, 1 will cost 28p. Hence 12 drinks will cost 12 × 28p, which is £3.36.

WORKED EXAMPLE 8

Stanley was appointed to paint the bridges in his home town. He found that on average he managed to paint 18 feet of bridge in a 6-hour day. How long would it take him to paint the longest bridge in his town, which was 68 feet?

We can find a multiplying connection between feet and hours by saying he will paint 3 feet in an hour, or that 1 foot will take 20 minutes. If 1 foot takes 20 minutes then 68 feet will take 68 × 20 minutes = 1360 minutes. Divide by 60 to change to hours and this rounds off to 23 hours. Since he works 6 hours a day it will take him 3 days and 5 hours. (He'll probably say it takes him 4 days!)

EXERCISE 5

If the cost of washing a wall 2 m long is 75p, what would you expect to be the cost of washing a similar wall 3 m long?

INVERSE PROPORTION

Inverse proportion is when there is a dividing connection between two things. So, this time, as one gets bigger the other gets smaller.

WORKED EXAMPLE 9

If I drive from home to work at an average speed of just 30 miles per hour it takes me 40 minutes. How long would it take me if I drove at 50 miles per hour?

Clearly, as the speed gets bigger, the time will get smaller. This time we would say that if at 30 mph I take 40 minutes, then at 1 mph I would take 30 × 40 = 1200 minutes. So if I drive at 50 mph I will take 1200 ÷ 50 = 24 minutes.

EXERCISE 6

Five men can assemble a car in 6 hours. How long will it take seven men?

DIRECT OR INVERSE?

So, when you meet a problem that involves proportion as we've described here, you need to think carefully whether it is direct or inverse so that you approach the problem the right way up!

EXERCISE 7

In an average 4-hour evening at her fish 'n' chip shop, Auntie Beatie will serve 72 portions of fish and chips. She decides to try opening her shop for 5 hours. How many portions of fish and chips would she expect to sell during the 5 hours?

5 ▷ RATE

We use the idea of *rate* quite a lot, from speed, which is the rate of distance travelled per unit of time, to costs of, say, a tennis court, which may be at the rate of 30p per half hour. We shall look at various examples which use the idea of a rate as a description of how something is changing.

SPEED

Speed is the rate of change of distance. For example, if you are walking at 5 miles per hour we mean just that. Every hour you walk you will have covered 5 miles, so in 2 hours you will have walked 10 miles, in 3 hours you will have walked 15 miles and so on.

If we know the distance travelled and the time taken to do it, then we can find the speed by dividing distance by time, being careful with the units.

It is useful to remember that

$$\text{speed} = (\text{distance travelled}) \div (\text{time taken}).$$

WORKED EXAMPLE 10

The 0815 train left Sheffield and arrived in Penzance, 370 miles away, at 1315. What was the average speed of the train?

The distance is 370 miles, and the time is from 0815 to 1315, which is 5 hours. Hence the average speed is $370 \div 5 = 74$ miles per hour.

EXERCISE 8

The train in the previous example stopped at Birmingham, 87 miles from Sheffield. At approximately what time would it stop there?

THE POUND ABROAD

AUSTRIA	Sch21.90
BELGIUM	Fr65.10
CANADA	C$2.030
DENMARK	DKr11.750
FRANCE	Fr10.100
GERMANY	DM3.1300
GREECE	Drc203.00
HOLLAND	Gld3.5300
HONG KONG	HK$11.400
IRELAND	I£1.0435
ITALY	L2138.00
JAPAN	Y229.00
NORWAY	NKr10.950
PORTUGAL	Esc213.00
SPAIN	Pes200.00
SWEDEN	SKr10.350
SWITZERLAND	Fr2.520
USA	$1.4700

EXCHANGE RATES

Shown in the table are the exchange rates for the British pound in 1985. The table indicates the amount of each foreign currency you would have received for £1. As these rates change slightly day by day, the table is obviously out of date; but, nevertheless, it will illustrate the point.

If you were going to France on holiday and wanted to take £50 with you, the money would not be much use in France unless you exchanged it for their currency, francs. The exchange rate given is 10.1 francs to the £1. So £50 will be exchanged for 50×10.1 francs, which is 505 francs. To exchange back again we need to divide by the exchange rate.

WORKED EXAMPLE 11

Freda came back from Switzerland with 68.6 francs. Using the table above, how much could she exchange this for in Britain?

The exchange rate is 2.52 francs to the £1. So 68.6 francs will be exchanged for $68.6 \div 2.52 = £27.22$ (when rounded off).

EXERCISE 9

Gary went on holiday to Sweden, exchanging £105 into Swedish kronor. Using the exchange rate table above, how much would he get?

6 ⟩ **SIMILAR SHAPES** Two shapes are said to be *similar* if all their corresponding angles are equal and the ratios of the corresponding lengths are also equal. For example:

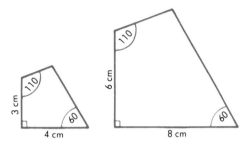

All the corresponding angles you can see are equal and the ratio of each pair of corresponding sides is 1 : 2.

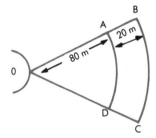

The diagram shows a 'javelin field' *OABCD*. The arc length *AD* is 50 metres. How long is the arc length *BC*?

We have two similar sectors here, *OAD* and *OBC*. Draw them separately if it will help you. The ratio of the lengths of *OAD* to *OBC* is 80 cm : 100 cm, which will simplify to 4 : 5. Hence the ratio of *AD* to *BC* is also 4 : 5, which gives 50 : *BC* = 4 : 5. If we rewrite these ratios as fractions the problem is simplified to $\dfrac{BC}{50} = \dfrac{5}{4}$ which then gives us $BC = \dfrac{5 \times 50}{4} = 62.5$ metres.

RATIOS OF SIMILAR SHAPES

Consider the following two shapes.

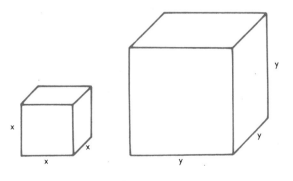

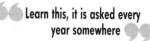

Both are cubes and hence have similar shapes, but consider their ratios:

> Length ratio is $x : y$
> Area ratio is $x^2 : y^2$ if you consider any face.
> Volume ratio is $x^3 : y^3$

This special relationship always exists between ratios of length, area and volume of similar shapes.

WORKED EXAMPLE 13

If you doubled the dimensions of a 1 pint milk bottle, what would happen to (a) the area of the milk bottle top, and (b) the volume?

If the ratio of the length is $1 : 2$, then the ratio of areas will be $1^2 : 2^2$, i.e. $1 : 4$, so the milk bottle top will be an area 4 times that of the original. The ratio of the volume will be $1^3 : 2^3$, i.e. $1 : 8$, so the volume would be 8 times bigger – that is, a 1 gallon bottle.

WORKED EXAMPLE 14

A gold statue of height 3 metres was melted down. The gold was then cast into 64 small similar statues. What is the height of each small statue?

The ratio of volume will be $64 : 1$, so to find the ratio of the lengths we need to find the cube root of each number, to give $4 : 1$. Therefore, each small statue will be of height 3 m $\div$ 4 which is 0.75 m or 75 cm.

EXERCISE 10

For a stage production of Jack and the Beanstalk, all the giant's furniture was larger than ordinary furniture but of a similar shape. For example, his coffee table, which was 270 cm long, was copied from a real coffee table 90 cm long. How much heavier should his portable TV have been than an ordinary one, and do you think the dressed up giant would be able to carry it?

SOLUTIONS TO EXERCISES

S1

(a) Since 3 eggs will make a cake sufficient for 12 people, 1 egg will be enough for 4 people, and 2 eggs will be enough for 8 people.

(b) Each ingredient will require $\frac{2}{3}$ of the given recipe, but we need to round off to sensible proportions. Therefore the actual $\frac{2}{3}$ recipe of 2 eggs, 133.3 g of flour, 66.67 g of margarine, 100 g of sugar and $\frac{2}{3}$ of a pinch of salt will in practice be 2 eggs, 130 g flour, 70 g margarine, 100 g sugar and 1 pinch of salt.

S2

The ratio of the investment is $900 : 500$, which is the same ratio as $9 : 5$. Hence the profit of £12 000. needs to be divided into 14 parts, 9 parts for Anna and 5 parts for Beth. Each part is £12 000 $\div$ 14 = £857.14 (rounded off). So Anna will receive £857.14(. . .) $\times$ 9 = £7714.29 (it is important to use the exact value of 12 000 $\div$ 14 here for accuracy, but then round off). Similarly Beth will receive £857.14(. . .) $\times$ 5 = £4285.71.

S3

The shortest route is from Blueport, through Toxintown then on to Castlebury. Using a ruler and measuring roughly round the roads gives 4 cm from Blueport to Toxintown and then 10 cm from Toxintown to Castlebury, a total of 14 cm. With a scale of 1 : 200 000, the actual road distance will be $14 \times 200\,000$ cm $= 2\,800\,000$ cm. Divide by 100 to change to 28 000 metres, then divide by 1000 to change to 28 kilometres.

S4

If Tom's 12 m² needs 1500 ml of 'Weedo', then 1 m² will need $1500 \div 12 = 125$ ml. Hence Ken's 20 m² will need 125 ml $\times 20 = 2500$ ml. The cheapest way for Ken to buy his 'Weedo' would be one 1500 ml bottle and one 1 litre bottle.

S5

$$\text{Cost} = \frac{3}{2} \times 75\text{p} = £1.13$$

S6

We can see that more men will take less time, and so if we consider 5 men assembling a car in 6 hours, then 1 man will take $5 \times 6 = 30$ hours. So 7 men will take $30 \div 7 = 4.28$ hours, which is about 4 hours 20 minutes.

Note: To calculate $30 \div 7$ exactly your calculator will give you 4.285 714 3. We know the 4 is 4 hours, but what about the 0.285 714 3? This is a decimal fraction of the hour, and to change it to minutes you need to first subtract the 4 in your calculator to give the decimal fraction part only of 0.285 714 3. Now multiply this by 60 to give you 17.142 857 which rounds off to 17 minutes.

S7

During 1 hour Auntie Beatie serves $72 \div 4 = 18$ portions of fish and chips, so during 5 hours she would expect to serve $18 \times 5 = 90$ portions.

S8

The train is travelling at 74 miles per hour so will cover 74 miles every hour. So divide 87 by 74 to see how many hours it takes to travel this distance – this is 1.175 675 7 hours. You may remember we've met this problem before (see S6). We have 1 whole hour, so take 1 away in the calculator to leave 0.175 675 7, multiply by 60 to get 10.540 541 which rounds off to 11 minutes. Therefore the train takes 1 hour 11 minutes. Add this time on to 0815 and you get 0926, the approximate time of arrival at Birmingham.

S9

Multiply 105 by the exchange rate of SKr 10.35 to give SKr 1086.75 which is the amount Gary would take on holiday.

S10

Ratio of lengths is 270 : 90, which cancels to 3 : 1. Hence the ratio of the weights, which is the same as the ratio of volumes, will be $3^3 : 1^3$ which is 27 : 1. So the portable TV would have to be 27 times heavier, far too heavy to be carried about.

E X A M   T Y P E   Q U E S T I O N S

Q1

A tape was being recorded for a disco where they wanted a mixture of 'pop' and 'heavy metal' in the ratio of 4 to 1. If the tape lasts for 2 hours, how long will be allocated to each type of music?

Q2

A multi-storey car park takes two hours to fill at the rate of 9 cars per minute. How long would it take to fill at the rate of 6 cars per minute?   (SEG)

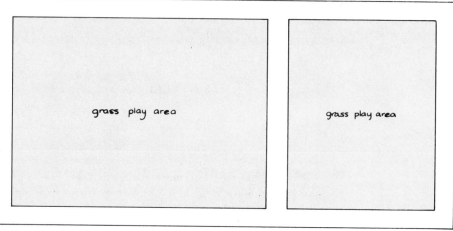

Q3

(a) The Fig. above is a plan of a park. Sarah takes $3\frac{1}{2}$ hours to mow the larger grass play area. How long will it take her to mow the smaller grass area?

(b) There are railings all round this park which Andrew, Charles and Dianne paint between them in 7 hours. How long would it have taken had Sarah been helping them also?

Q4

SCONES

8 oz plain flour
1 teaspoonful salt
1 teaspoonful bicarbonate of soda
2 teaspoons cream of tartar
$1\frac{1}{2}$ oz margarine
About $\frac{1}{4}$ pint milk

This recipe is enough for 8 scones.

(i) How much flour is needed for 12 scones?

(ii) How much milk is needed for 12 scones?

(NEA)

Q5

At a scout jamboree in London, Billy from Belgium came with 2000 francs, Gerry from Germany came with 100 Deutschmarks, and Olly from Holland came with 120 guilders. When their money was exchanged (use the table given on page 68), who had the most and who had the least?

Q6

100 g Price 97 p

80 g Price 73 p

50 g Price 47 p

Whiteside toothpaste is sold in three sizes.
Which size is the best value for money?
Describe how you reached your answer.

(SEG)

Q7

A model was made of the village of Banner Cross with a scale of 1 : 30.

(a) How wide would each model road be if the actual roads were 3 metres wide?

(b) The model village hall was 80 cm long, so how long would the actual village hall be?

Q8

The top part A of a cone is removed to leave the lower part B. The vertical heights of A and B are equal. If the volume of A is 3.14 cm³, what is the volume of the whole cone?   (SEG)

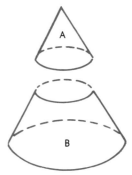

Q9

The scales show that 9 marbles balance with 6 balls. Two balls are removed. How many marbles must be removed so that the scales will balance?   (NEA)

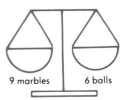

9 marbles   6 balls

Q10

The two cubes illustrated are gift boxes of a similar shape, having lengths in the ratio 5 : 9.

(a) The smaller box requires 250 cm² of wrapping paper. How much will the larger box need?

(b) The volume of the larger box is 1000 cm³. What will be the volume of the small box?

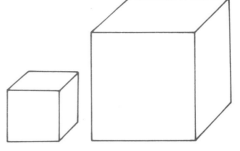

OUTLINE ANSWERS TO EXAM QUESTIONS

A1

The tape lasting 120 minutes needs to be divided into five parts, giving four for the 'pop' and one for the 'heavy metal'. So each part will be $120 \div 5 = 24$ minutes. Hence 'pop' will be allocated $24 \times 4 = 96$ minutes and 'heavy metal' 24 minutes.

A2

(2×9) hours at a rate of 1 car per minute, hence $(2 \times 9) \div 6$ at a rate of 6 cars per minute = 3 hours.

A3

(a) You need to use a ruler and measure the sides of each grass area. The area on the plan of the large grass area is 5 cm × 7 cm = 35 cm². So Sarah mows 35 m² in $3\frac{1}{2}$ hours, which is 210 minutes. She will therefore mow 1 m² in 210 ÷ 35 = 6 minutes. The size of the small grass area is 4 cm × 5 cm = 20 cm². Hence the time taken to mow the 20 m² area will be 20 × 6 = 120 minutes, which is 2 hours.

(b) If three people paint the fence in 7 hours, then one person will paint it in 21 hours. Hence if Sarah helps out, the four of them will paint the fence in 21 ÷ 4 = 5.25 hours, which is $5\frac{1}{4}$ hours.

A4

Ratio of menus is 8 : 12 or 2 : 3; hence

(i) 12 oz

(ii) $\frac{1}{4} \times \frac{3}{2} = \frac{3}{8}$ pint.

A5

Billy came to London with 2000 ÷ 65.1 = £30.72, Gerry came with 100 ÷ 3.13 = £31.95 and Olly with 120 ÷ 3.53 = £33.99. So Olly came with most, Billy the least.

A6

Calculations pence per gram:

Family: $\frac{97}{100}$ = 0.97 pence per gram.

Large: $\frac{73}{80}$ = 0.9125 pence per gram.

Standard: $\frac{47}{50}$ = 0.94 pence per gram.

Best value: Large. The answer would only be accepted if the calculations were shown.

A7

(a) The model road will be $\frac{1}{30}$ of the actual road. Hence a width of 300 cm ÷ 30 gives 10 cm for the model road.

(b) The actual village hall will be 30 times longer than the model, hence 30 × 80 cm gives us 2400 cm or 24 metres.

A8

Ratio of heights for top cone : whole cone = 1 : 2.

Hence

ratio of volumes   = 1 : 8

volume of whole cone   = 8 × 3.14 = 25.12 cm³.

A9

They balance in the ratio of 9 : 6, which is 3 : 2. Hence, if 2 balls taken, this balances with 3 marbles taken.

A10

The ratio of the lengths is 5 : 9, so for (a), where we need to consider area, the ratio of areas will be $5^2 : 9^2$, which is 25 : 81. Writing this as fractions you should have $\frac{(\text{big cube})}{250} = \frac{81}{25}$ which gives big cube $= \frac{(81 \times 250)}{25} = 810$ cm². So the area of wrapping paper needed for the big cube is 810 cm².

(b) We need here to consider volumes, and the ratio of volumes is $5^3 : 9^3$ which is 125 : 729. You can now work out the volume of the small box by $\frac{(\text{little cube})}{1000} = \frac{125}{729}$, which is calculated as little cube $= \frac{(125 \times 1000)}{729} = 171.5$ cm³, the volume of the small box.

G R A D E   C H E C K L I S T

FOR A GRADE F

You should know:

> what is meant by direct and inverse proportion;
>
> what exchange rates are;

and be able to:

> perform straightforward calculations involving average speed;
>
> use ratio, direct and inverse proportions in simple applications;
>
> make use of measures of rate in everyday use, e.g. cost per hour, miles per gallon, in the solution of simple problems;
>
> use exchange rates to convert to and from the British pound.

FOR A GRADE C

You should also be able to:

> use simple ratio and proportion, such as 3 : 2, 1 : 20, for example to calculate ingredients for recipes;
>
> identify best buys from given data;
>
> obtain dimensions from scale drawings, models and maps involving scales such as 1 : 5, 1 : 50, 1 : 25 000.

FOR A GRADE A

You should also be able to:

> relate scale factors to situations in both two and three dimensions;
>
> calculate actual lengths, areas and volumes from scale models.

STUDENT'S ANSWER - EXAMINER'S COMMENTS

QUESTION

A sealed hollow cone with vertex downwards is partially filled with water. The volume of water is 250 cm^3 and the depth of the water is 60mm. Find the volume of the water which must be added to increase the depth to 80 mm.

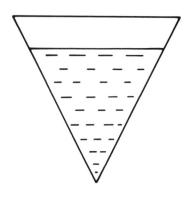

Answer

66 **good explanation of what is happening** 99

USING SIMILAR SHAPES:

RATIO OF LENGTH IS 60:80

RATIO OF VOLUME IS 250 : x

$= 60^3 : 80^3$

SO $\dfrac{x}{250} = \dfrac{80^3}{60^3}$

66 **shame, the final question has not been answered. To answer the question you must subtract 250 from this. Still, the method will get you high marks for this answer.** 99

$x = \dfrac{80^3 \times 250}{60^3} = 592$

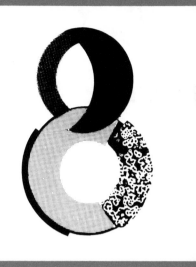

ALGEBRA

G E T T I N G S T A R T E D

All the GCSE courses have a common aim for *Algebra*. This is that the emphasis within algebra should be on the use of letters or words to *communicate mathematical information* and not simply to solve equations. If you intend to go any further in mathematics after this course, it is essential that you appreciate the use of algebra. You need to be confident in your ability to represent general situations in algebraic terms and then to manipulate those terms.

USEFUL DEFINITIONS

Constant	not changing, as in $y = x + 5$, the 5 is always 5
Domain	the set of numbers from which a mapping starts
Expand	to multiply out the brackets
Factorise	split into expressions that multiply together to make the whole
Function	an algebraic 'happening'; a rule for changing numbers
Generalise	express in general terms, usually using an algebraic formula
Index	the raised figure that gives the power, e.g. the 3 in y^3
Linear	an expression involving single variables of power 1, e.g. $x + y = 3$
Quadratic	an expression involving a squared term, e.g. $3x^2 - 5x = 7$
Range	the set of numbers to which a mapping maps
Simplify	to make easier; in this chapter it will mean to multiply out brackets and collect like terms
Simultaneous	at the same time
Solve	to find the numerical value of the letter
Transposition	to change the subject of a formula or equation
Variable	a letter which may stand for various numbers

FORMULAE
EQUATIONS
ALGEBRAIC SHORTHAND
EXPANSION
FACTORISATION
SIMPLIFICATION
QUADRATIC EQUATIONS
SIMULTANEOUS EQUATIONS
FRACTIONAL EQUATIONS
INEQUATIONS
INDICES
FUNCTIONS
APPLICATION

E S S E N T I A L   P R I N C I P L E S

1 **FORMULAE**   If you can follow through a *flowchart* like this

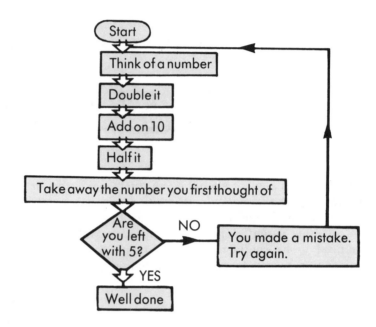

then you can easily substitute numbers into simple formulae which are written out. For example, Gladys, the office secretary, paid wages to the workers using the formula.

wage = £40 + £10 multiplied by the number of years worked.

So to find Kevin's wage when he has worked for 15 years we need to substitute 15 into the formula. This will then give Kevin the wage of £40 + £10 × 15 = £40 + £150 = £190. This example could have been done with a flowchart as

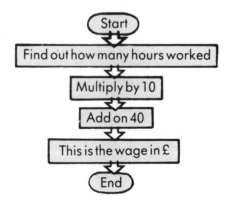

Sometimes the formula is not in words and not in a flowchart, but in what are called 'algebraic terms'. For example, the area of a rectangle is given by the formula

$A = bl$

where A = area, b = breadth, l = length.

For this type of formula we need to remember the basic rules of algebra. For example,

$3t$	means	3 multiplied by t
xy	means	x multiplied by y
$\dfrac{p}{3}$	means	p divided by 3

WORKED EXAMPLE 1

The cost of a pot of coffee in a cafe is calculated by using the formula: $C = 25 + 15n$, where C is the cost of the pot of coffee in pence and n is the number of people sharing the pot of coffee. Calculate the cost of a pot of coffee for 6 people.

We substitute $n = 6$ into the formula $C = 25 + 15n$. The $15n$ means multiply 15×6 in this case, which is 90. Hence the cost is given by $C = 25 + 90$, which is 115p or £1.15.

EXERCISE 1

A bank pays car expenses to its inspectors when they travel to various branches around the country. The expenses work out as follows.

For journeys of 50 miles or less: Amount $= £\dfrac{24N}{100}$ and

for journeys of more than 50 miles: Amount $= £12 + \dfrac{£20(N - 50)}{100}$

where N is the number of miles travelled.

How much will be paid for a journey of (i) 30 miles, (ii) 80 miles?

BRACKETS

We often need to make sure that in a formula certain numbers are calculated first. We do this by the use of brackets. For example, in the formula $A = (d + e) \div 2$, it is important to add together d and e before dividing the answer by 2, or else you will get quite a different number. So if a bracket appears in a formula, work out the bracket first.

For example, $9 - (5 - 2)$ is equal to $9 - 3$, which is 6, whereas without the bracket this would be read from left to right and be $9 - 5 - 2$ which is $4 - 2$, giving us 2.

BODMAS

This brings us to what do we do when there are no brackets to indicate what to do first. Do we always work from left to right, or is there some other rule? The answer is that if we follow the rule of BODMAS this gives us the order. BODMAS stands for the phrase

Brackets, Of, Division, Multiplication, Addition, Subtraction

and we should do the things in that order. For example, the horrible sum of

$$10 \div 2 \ + 8 \times 3 \ - \tfrac{1}{2} \text{ of } 6 + (4 - 2)$$

is done like this:

Brackets	$10 \div 2 \ +$	$8 \times 3 - \tfrac{1}{2}$	of 6 +		2
Of	$10 \div 2 \ +$	$8 \times 3 -$	3	+	2
Division	$5 \quad +$	$8 \times 3 -$	3	+	2
Multiplication	$5 \quad +$	$24 \quad -$	3	+	2
Addition		29	$-$	5	
Subtraction		24			

If any two or more of the same signs are next to each other we work from left to right. For example, $10 - 6 - 2$ will be $4 - 2$ which is 2.

EXERCISE 2

Evaluate $\tfrac{1}{2}$ of $12 \div 6 + 2 \times (3 - 1) - 4$.

TRANSPOSITION

It is often necessary to be able to change a formula round to help you find a particular piece of information. For example, if I used the formula $C = \pi D$, where C is the circumference of a circle and D is the diameter, then to find the diameter when the circumference is 100 cm, it would be more convenient to change the formula round to give $D = \dfrac{C}{\pi}$. I then have only a simple substitution to do. This changing round of formulae is called *transposition* of formulae and what we are doing is changing the subject of a formula. The subject of a formula is the single letter (or word) usually on the left-hand side all by itself. For example, t is the subject of the formula $t = \dfrac{d}{v}$.

Here are some rules for changing the subjects of formulae:

> **This is the very basis of most of your algebra, Understand this and you're almost home!**

1   You can move any letter (or word) from one side of the equation to the other as long as it is operating on *all* the rest of that side. For example, in the formula $v = u + 6t$, the u can be moved since it is adding to the rest of that side, but the t cannot be moved yet as it is only multiplying the 6.

This simple list of formulae should help you see when we can move terms:

$v = u + 6t$ we could move either the u or the $6t$

$A = lb$ we could move either the l or the b

wage = hours × hourly rate   we could move either hours or hourly rate

$t = \dfrac{d}{v}$ we could move either the d or the v

$x = \dfrac{y + 1}{7}$ we could move either the 7 or the $(y + 1)$

$w = n(y - 10)$ we could move either the n or the $(y - 10)$

2   When a letter (or word) has been moved from one side to the other, it does the 'opposite thing' to the other side. For example, if something was added, then when it moved it would be subtracted, or if something was multiplied then when it moved it would divide. The following examples will help to illustrate these points.

$v = u + 6t$
    can be changd to $v - u = 6t$ or $v - 6t = u$

$A = lb$
    can be changed to $\dfrac{A}{l} = b$ or $\dfrac{A}{b} = l$

wage = hours × hourly rate
    can be changed to $\dfrac{\text{wage}}{\text{hours}} = \text{hourly rate}$ or $\dfrac{\text{wage}}{\text{hourly rate}} = \text{hours}$

$t = \dfrac{d}{v}$
    can be changed to $tv = d$ or $\dfrac{t}{d} = \dfrac{1}{v}$

$y = 6x - 10$
    can be changed to $y + 10 = 6x$ (or $y - 6x = -10$)

WORKED EXAMPLE 2

A firm calculates wages using the formula

    wage = £15 + hours × hourly rate

Change the formula to make hours the subject, and hence find the number of hours that Philip will have to work to earn £180 when his hourly rate is £5.50.

From wage = 15 + hours × hourly rate we change to
wage −15 = hours × hourly rate, which we then change to
(wage − 15)/hourly rate = hours. Now, substitute wage = 180 and hourly rate = 5.50, to give hours equal to 30.

EXERCISE 3

You are given the simple interest formula

$$I = \frac{PRT}{100},$$

where I = interest, P = the principal amount, R = rate of interest, T = time.

Change the formula to make R the subject, and hence find the rate that gives an interest of £6 when the principal amount is £20 and the time is 4 years.

2 ▷ EQUATIONS   *Equations* are mathematical statements with letters in place of numbers, but most importantly they contain an equals sign. *Formulae* are equations which can contain any number of letters in place of numbers, as you have already seen.

There are many different types of equations that you will come across, and when you are asked to solve an equation you are expected to find the value of the letter (or letters) that make the mathematical statement true. For example, $50 = 16y + 2$ is an equation, and if we were to solve it we would find that $y = 3$, since that is the only value of y that will make both sides of the equation equal.

LINEAR EQUATIONS

Linear equations are equations that involve *single* variables of power 1. They contain no expressions such as x^2, y^3, $\frac{1}{x}$, xy, etc. Some examples of linear equations of the type you will meet are:

$$x + y = 10, \quad C = D, \quad W = 50 + 10N.$$

We move numbers around in an equation in exactly the same way as we moved letters around in a formula.

WORKED EXAMPLE 3

Solve the equation $18 = 5x + 7$.

As before, we can change $18 = 5x + 7$ into $18 - 7 = 5x$ which is $11 = 5x$, which then becomes $\frac{11}{5} = x$. Hence $x = 2.2$.

So remember, the rules applying to formula also apply to equations and can be summarised as:

'If it's doing it to all the rest, you may move it and make it do the opposite to the other side.'

EXERCISE 4

The circumference of a wheel is approximately 3.1 multiplied by its diameter. What will be the diameter of a bicycle wheel with circumference 56 inches?

3 ▷ ALGEBRAIC SHORTHAND   You should be familiar with the shorthand we use in algebra. For example, ab meaning a multiplied by b, $\frac{3}{x}$ meaning 3 divided by x, and combining similar terms as $5x + 3x$ to give $8x$.

There are other very common things that we have a special shorthand for, and those are *indices*. For example, we use the shorthand a^4 to mean $a \times a \times a \times a$.

We also have a very easy way to multiply and divide two similar index terms which should be remembered.

$$a^x \times a^y = a^{x+y}; \text{for example } 2^7 \times 2^5 = 2^{12}$$
$$a^x \div a^y = a^{x-y}; \text{for example } y^6 \div y^3 = y^3.$$

You also need to know that for any number x, then $x^1 = x$ and $x^0 = 1$. We also use negative indices to show numbers of a fractional form. For example,

$$x^{-2} = \frac{1}{x^2} \text{ and } 5^{-1} = \frac{1}{5}.$$

EXERCISE 5

Write down the value of (i) $6^2 \times 6^4$, (ii) $x^5 \div x^6$

4 ▷ EXPANSION

We first met brackets when we were told to work out the brackets first. In algebra we extend the use of brackets to keep terms together or to factorise.

But first let us look at *expansion of brackets*. This usually means 'multiply out'. For example, $6(x + 3)$ would be expanded to $6x + 18$, so you see why it is sometimes called 'multiply out'.

WORKED EXAMPLE 4

Solve the equation $6x = 4(x + 5)$.

Expand the bracket first to give $6x = 4x + 20$. Then we can solve in the same way as before to give $2x = 20$ then $x = 10$.

WORKED EXAMPLE 5

Expand $x(x + y)$.

This will 'multiply out' to give $x^2 + xy$.

MULTIPLYING BRACKETS

The most difficult type of bracket expansion you will meet *at all but the highest level* is that of the type $(x + a)(x + b)$, where everything in the first bracket has to multiply everything in the second bracket. We can illustrate this best with a diagram:

$$(x + 6)(x + 4) = x^2 + 4x + 6x + 24$$
$$= x^2 + 10x + 24$$

EXERCISE 6

Expand: (i) $(x + 2)(x + 3)$; (ii) $(t - 1)(t + 2)$

5 ▷ FACTORISATION

You are also expected to be able to *factorise* a simple algebraic expression. For example, by looking at the formula $A = \pi r^2 + \pi d$ we notice that on the right-hand side both terms contain π, and so this can be rewritten as $A = \pi(r^2 + d)$. Some more examples to illustrate this are:

$$P = 2l + 2b \text{ can be factorised to } P = 2(l + b)$$
$$D = \pi r - r^3 \text{ can be factorised to } D = r(\pi - r^2)$$

The way to check your answer is to expand the bracket out again to see if you get the same as you started with.

EXERCISE 7

Factorise: (i) $A = \pi R^2 - \pi r^2$; (ii) $T = 5pq + 10p^2$.

6 ▷ SIMPLIFICATION

This means what is says! It is what we do to make an expression more simple. For example, to simplify $t = 3(x + y) + 4(2x + y)$ we would expand both brackets to give $t = 3x + 3y + 8x + 4y$, which will then simplify to $t = 11x + 7y$ by combining similar terms.

EXERCISE 8

Simplify: (i) $y = 4(x + t) + 2(4x - t)$; (ii) $p = 3(x - y) - 2(3x - y)$.

FURTHER EXPANSION

You should remember from the previous sections how to expand $(x + a)(x + b)$. Well, in exactly the same way at the higher levels you are expected to be able to expand $(ax + b)(bx + c)$. For example:

$$(3x + 2)(4x + 7) = 12x^2 + 21x + 8x + 14$$
$$= 12x^2 + 29x + 14.$$

QUADRATIC EXPRESSION

A quadratic equation is an expression where the highest power involved is a 2 and no negative powers are used. For example, $3x^2 + 6x - 7$ or $5x^2 - 2y^2$

QUADRATIC FACTORISATION

Factorisation means putting a quadratic back into its brackets if at all possible, although in an examination situation you are not likely to be asked to factorise a quadratic that will not do so.

It is helpful when factorising to first consider what the signs may be.

> Relax; if you've been asked to factorise then it will work out if you follow these simple rules.

1   When the last sign in the quadratic $ax^2 + bx + c$ is positive, both signs in the brackets are the same as the first sign in the quadratic. For example,

$$6x^2 + 7x + 2 = (\quad + \quad)(\quad + \quad)$$

and

$$6x^2 - 7x + 2 = (\quad - \quad)(\quad - \quad)$$

2   When the last sign in the quadratic $ax^2 + bx - c$ is negative, the signs in the brackets are different. For example,

$$6x^2 + x - 2 = (\quad + \quad)(\quad - \quad)$$

Once you've sorted out the signs then you need to look at the numbers. Follow through these examples for the general way to do this.

WORKED EXAMPLE 6

Factorise $6x^2 + 7x + 2$.

By looking at the signs we see that the brackets both contain a '+', so $6x^2 + 7x + 2 = (\quad + \quad)(\quad + \quad)$. We see that the end numbers in each bracket must multiply to give 2, and the only way to do this is to have 2×1. Hence $6x^2 + 7x + 2 = (\quad + 2)(\quad + 1)$.

Now we see that the first numbers in each bracket must multiply to give 6, and we could have 6×1 or 3×2, but the combination we need is to multiply with the 2 and the 1, so that their sum is 7. We ask ourselves which of

$$\begin{Bmatrix} 3 \times 2 \\ 2 \times 1 \end{Bmatrix} \begin{Bmatrix} 2 \times 2 \\ 3 \times 1 \end{Bmatrix} \begin{Bmatrix} 6 \times 2 \\ 1 \times 1 \end{Bmatrix} \text{ or } \begin{Bmatrix} 1 \times 2 \\ 6 \times 1 \end{Bmatrix}$$

give a combined total of 7, and we see that the only one which does is

$$\begin{Bmatrix} 2 \times 2 \\ 3 \times 1 \end{Bmatrix},$$

so the factorisation is $(3x + 2)(2x + 1)$.

WORKED EXAMPLE 7

Factorise $2x^2 + 5x - 3$

We factorise by looking at the signs first and noticing that both signs will be different, hence $(\quad + \quad)(\quad - \quad)$. The -3 indicates we need a 3 and a 1 at the end of each bracket to give $(\quad + 3)(\quad - 1)$ or $(\quad + 1)(\quad - 3)$. Now, a product of 2, i.e. 2 and 1, combining with 3 and 1 in such a way to give a difference of 5, will give us $(x + 3)(2x - 1)$.

EXERCISE 9
Factorise: (i) $x^2 + 8x + 12$; (ii) $2x^2 + 5x - 3$.

DIFFERENCE OF TWO SQUARES
If you expand $(x + y)(x - y)$ in the usual way you get $x^2 - xy + xy - y^2$, which is $x^2 - y^2$. This is called the *difference of two squares*. It is very valuable to be able to recognise this situation either way, so that when you are faced with any example of the type, say, $9x^2 - 25$, you recognise both terms are squares and you can write down

$$9x^2 - 25 = (3x + 5)(3x - 5).$$

EXERCISE 10
Factorise: (i) $x^2 - 16$; (ii) $4y^2 - 9$; (iii) $\pi(R^2 - r^2)$.

7 ▷ QUADRATIC EQUATIONS

A quadratic equation is an equation involving a *quadratic* expression, e.g. $2x^2 - 3x + 6 = 0$. To solve a quadratic equation, for example $x^2 + x - 6 = 0$, we initially need to factorise to give $(x + 3)(x - 2) = 0$. Then since the product of the two numbers is zero, one bracket or both must be equal to zero. In this case either $x + 3 = 0$ or $x - 2 = 0$, and we have two simple linear equations which solve to $x = -3$ or $x = 2$.

WORKED EXAMPLE 8

❝ You may have learnt the formula method, if so the formula will be given to you on a formula sheet and you simply substitute into it. ❞

Solve the equation $2x^2 + 5x + 4 = 7$.

You must remove the 7 to make the quadratic equal zero. This will give us $2x^2 + 5x - 3 = 0$, which factorises to $(2x - 1)(x + 3) = 0$. This gives $2x - 1 = 0$ or $x + 3 = 0$. Therefore the solution is $x = \frac{1}{2}$ or $x = -3$.

EXERCISE 11
Solve: (i) $y^2 + y - 6 = 0$; (ii) $16x^2 - 9 = 0$; (iii) $x^2 + x = 20$.

8 ▷ SIMULTANEOUS EQUATIONS

Simultaneous equations are where we have a pair of linear equations that both need solving at the same time (hence simultaneous). The technique is initially to eliminate one variable to find a solution to the other one, and then to substitute the found variable to complete the solution.

WORKED EXAMPLE 9

Solve the simultaneous equations $4x + y = 14$ ...(1)
                                      $2x + y = 8$ ...(2)

To eliminate one of the variables, subtract the equations–bottom from top–which will give

$$2x = 6$$

so $x = 3$. Now, we substitute this answer in the simplest equation ((2) above) to give $6 + y = 8$, so $y = 2$. Therefore the solution of the simultaneous equation is $x = 3$ and $y = 2$. *NB:* You should always check this solution and see that it works with the other equation.

WORKED EXAMPLE 10

Solve the simultaneous equations $4x - y = 3$ ...(1)
                                      $3x + 2y = 16$ ...(2)

To eliminate y we need to double the whole of equation (1) to give

$$8x - 2y = 6$$
$$3x + 2y = 16$$

Now we can add the equations to eliminate y, to give $11x = 22$, making $x = 2$, which is substituted into equation (1) – being the simplest – to give $8 - y = 3$, or $y = 5$. So the final solution is $x = 2$, $y = 5$.

EXERCISE 12

Meg, with 36p, and Dudley, with 34p, go into a sweetshop and spend all their money on sweets. Meg buys 5 chews and 3 fudge bars, while Dudley buys 2 chews and 4 fudge bars. Find the price of chews and fudge bars.

9 ▷ FRACTIONAL EQUATIONS

Fractional equations of the type $\dfrac{x}{2} + \dfrac{x+1}{3} = 5$ can be solved by changing both fractions to equivalent fractions with the same bottom numbers, then carrying on using previous techniques.

WORKED EXAMPLE 11

Solve $\dfrac{x}{2} + \dfrac{x-1}{3} = 3$.

We shall change both fractions to have 6 on the bottom, i.e.

$$\frac{3x}{6} + \frac{2(x-1)}{6} = 3, \quad \text{giving} \quad \frac{3x + 2(x-1)}{6} = 3,$$

then $3x + 2x - 2 = 18$, which worked out gives $5x = 20$, so $x = 4$.

Be careful with the signs in this situation, and look carefully through the next example.

WORKED EXAMPLE 12

Solve $\dfrac{x}{4} - \dfrac{3(1-x)}{2} = 2$.

We can change the fractions to quarters, so $\dfrac{x}{4} - \dfrac{6(1-x)}{4} = 2$. This gives $\dfrac{x - 6(1-x)}{4} = 2$, so $\dfrac{x - 6 + 6x}{4} = 2$.

(Notice here that the -6 multiplies with $-x$ giving $+6x$.) The solution then will be $7x - 6 = 8$, giving $7x = 14$, or $x = 2$.

EXERCISE 13

Solve: (i) $\dfrac{x}{3} + \dfrac{x+4}{2} = 7$; (ii) $\dfrac{1}{x} + x = 2$.

10 ▷ INEQUATIONS

Equations with inequality signs in them are called *inequations* (but often referred to as inequalities also), and are solved in exactly the same way, except that when you have to multiply or divide *both* sides by a negative number the inequality sign turns round. For example, if we have $-2 < 5$, then $2 > -5$.

WORKED EXAMPLE 13

Find the range of values for which $5x + 3 > 4(x - 2)$.

Using normal equation techniques we can calculate $5x + 3 > 4(x - 2)$ to give $5x + 3 > 4x - 8$, so $x > -11$.

EXERCISE 14

Find the smallest integer which satisfies the inequality $6x - 8 > 3x + 9$.

11 ▷ INDICES

The use of indices can be extended to fractions to indicate roots. For example, $9^{1/2}$ will be $\sqrt{9}$ which is 3. This can easily be seen to be true because $9^{1/2} \times 9^{1/2} = 9^1 = 9$.

WORKED EXAMPLE 14

Which is the larger number, $27^{1/3}$ or $8^{2/3}$?

$27^{1/3}$ means $\sqrt[3]{27}$ (the cubed root of 27) which is 3.
$8^{2/3}$ means $8^{1/3} \times 8^{1/3}$ or $(8^{1/3})^2$, since $8^{1/3} = \sqrt[3]{8} = 2$, then $(8^{1/3})^2 = 2^2 = 4$.
Therefore $8^{2/3}$ is larger than $27^{1/3}$.

EXERCISE 15

Place the following numbers into order, the smallest first:

$$5^{-2}, \, 5^{1/2}, \, 5^0, \, 125^{-1/3}.$$

12 ▷ FUNCTIONS

This is another way of writing an algebraic formula. We generally use the notation $f(x)$ for the image of x under the function f. For example, where $f(x) = 2x^2 + 5$, then $f(3) = 23$, since substituting $x = 3$ into $2x^2 + 5$ gives us 23. In other words, it is another way of writing $y = 2x^2 + 5$ where we talk about $f(x)$ instead of y.

EXERCISE 16

Where $f(x) = (x - 1)^2$, what is (i) f(3), (ii) f(0), (iii) f(−3)?

DOMAIN AND RANGE

The domain of a function is its starting point, usually a set of numbers that the function can work on. The range is the set of values that the function is able to actually 'go to'.

For example, in the function $f(x) = x^2 + 3$, the domain can be the whole set of numbers, positive and negative, while the range is only the positive numbers equal to or larger than 3, since the lowest value of x^2 is zero.

COMPOSITE FUNCTIONS

These are functions made up of two (or more) simple functions. For example, when $f(x) = 2x$ and $g(x) = x^3$, then $gf(x)$, which is a composite function, is found by 'doing' $f(x)$ first, then $g(x)$.

WORKED EXAMPLE 15

Write down, in its simplest form, the composite function $fg(x)$ where $f(x) = x + 6$ and $g(x) = 5x$.

$fg(x)$ will be $f(g(x))$ which is $f(5x)$, giving $5x + 6$. Hence $fg(x) = 5x + 6$.

EXERCISE 17

Where $f(x) = 5x$ and $g(x) = \dfrac{1}{x}$, calculate the value of (i) fg(10), (ii) gf(10).

INVERSE

The inverse of a function is that function which returns numbers from the range back to the domain – it is the function 'the opposite way round'. The notation of an inverse function is usually $f^{-1}(x)$. For example, if $f(x) = 7x$, then $f^{-1}(x) = \dfrac{x}{7}$. To illustrate this we see that if $f(3) = 21$, then $f^{-1}(21) = 3$.

EXERCISE 18

Where $f(x) = x^2 - 1$, (i) state $f^{-1}(x)$, (ii) find $f^{-1}(3)$.

13 ＞   APPLICATION

You might use some of these ideas if you do any computer programming, since the letters we have used in this chapter can be put into computer programs as *variables*. They are called variables because they will vary, that is, be different at different times.

To put a formula into a computer program we need to establish which language we are using, as this could make a difference, but we shall assume that we are using BASIC. For example, to put the formula $t = \dfrac{d}{v}$ into a computer program we recognise that d is being divided by v, and so we would write a line: LET t = d/v (you see, it is BASICALLY the same).

Of course the computer needs to know what the letters 'd' and 'v' stand for, so we need to write a line: INPUT d, v, then print out the answer with: PRINT t. A nice little program that will do all this is:

```
10    PRINT "Tell me what d and v are"
20    INPUT d, v
30    LET t = d/v
40    PRINT "t is "; t
RUN
```

If you can, try this program out, then write your own using your own formulae and your own variables.

GENERALISATION

We also use algebra to help us show a pattern. For example, in the sequence 2, 4, 6, 8, . . ., we can see that the first number is 1×2, the second is 2×2, the third 3×2 and so on, so the nth number will be $n \times 2$ or $2n$. Hence this pattern can be described as $2n$ where $n = 1, 2, 3, 4, \ldots$.

WORKED EXAMPLE 16

Look at this number pattern:

		Row sum
1st row	1 = 1	$= 2^0$
2nd row	1 1 = 2	$= 2^1$
3rd row	1 2 1 = 4	$= 2^2$
4th row	1 3 3 1 = 8	$= 2^3$
5th row	1 4 6 4 1 = 16	$= 2^4$

Now, write down (a) the 6th row and row sum, (b) the 11th row sum.

(a) This is Pascal's triangle, and you should be able to see how the pattern builds itself down to give the 6th row as $1 + 5 + 10 + 10 + 5 + 1$, with a row sum of $32 = 2^5$.

(b) Look at the number of the row and the row sum, and you should see that the row sum of the nth row is 2^{n-1}. Hence the row sum of the 11th row will be 2^{11-1} which is 2^{10}. Now, 2^{10} is $2^5 \times 2^5$ which will be 32×32 which is 1024, i.e. the row sum will be $1024 = 2^{10}$.

FURTHER APPLICATIONS

You also need algebra when trying to describe an observed relationship. For example, a shop selling marbles had a notice saying how much you would have to pay to have your marbles polished!

Number of marbles	5	10	15	20	Other prices
Cost (pence)	15	25	35	45	on request

If you look carefully at the numbers you can find the simple formula that the shop is using to calculate its prices. As the number of marbles increases so too does the cost, but the shop is not simply adding each time nor just multiplying, so it looks like a combination of the two. By trial and error, you can work out that they double the number of marbles and add 5:

cost = 2 × number of marbles + 5.

(As you will see in Chapter 9, you could find this by drawing a graph.)

NOTE

It should be noted that within any GCSE course, the emphasis in algebra will be on the use of letters to express and identify relationships that you have found between sets of data, as in the two examples you have just seen. It will not be on the manipulation of arithmetical symbols (although some use of this is inevitable, especially at the highest levels).

S O L U T I O N S   T O   E X E R C I S E S

S1

(i) Use the first formula to give amount = $(24 \times 30) \div 100$, which is £7.20.

(ii) Use the second formula to give amount = $12 + 20(80 - 50) \div 100$, which is £18.

S2

1

S3

From $I = \dfrac{PRT}{100}$ we can change to $100I = PRT$, which we can then change to $\dfrac{100I}{PT} = R$. Hence the new formula is $R = \dfrac{100I}{PT}$. Substitute into this $I = 6$, $P = 20$ and $T = 4$, which will give you $R = 7.5$.

S4

If $C = 3.1 \times D$, then $D = \dfrac{C}{3.1}$; hence where $C = 56$, $D = \dfrac{56}{3.1} = 18.1$ (rounded off).

S5

(i) 6^6; (ii) $x^{-1} = \dfrac{1}{x}$.

S6

(i) $x^2 + 5x + 6$; (ii) $t^2 + t - 2$.

S7

(i) $A = \pi(R^2 - r^2)$; (ii) $T = 5p(q + 2p)$ (but see difference of two squares).

S8

(i) $y = 4x + 4t + 8x - 2t$
$\qquad = 12x + 2t = 2(6x + t)$

(ii) $p = 3x - 3y - 6x + 2y = -3x - y$.

S9

(i) $(x + 2)(x + 6)$; (ii) $(2x - 1)(x + 3)$.

S10

(i) $(x + 4)(x - 4)$; (ii) $(2y + 3)(2y - 3)$; (iii) $\pi(R + r)(R - r)$.

S11

(i) We can see that the signs will be different, so we need factors of 6 with a difference of 1 to give the answer of $(y + 3)(y - 2)$.

(ii) This is recognised as the difference of two squares; the square root of $16x^2$ is $4x$, the square root of 9 is 3, hence the solution is $(4x + 3)(4x - 3)$.

(iii) $x^2 + x - 20 = 0$
$\qquad (x + 5)(x - 4) = 0$
$\qquad x = -5$ and $+4$.

S12

Let the price of a chew be c, and the price of a fudge bar be f, and then we can write down a pair of simultaneous equations:

$\qquad 5c + 3f = 36$ (from Meg's information)

$\qquad 2c + 4f = 34$ (from Dudley's information).

Work these out and we get $c = 3$ and $f = 7$, so the price of a chew is 3p and the price of a fudge bar is 7p.

S13

(i) Use a common denominator of 6 to give

$$\frac{2x}{6} + \frac{3x + 12}{6} = 7, \quad \frac{2x + 3x + 12}{6} = 7.$$

Hence

$$5x + 12 = 42$$
$$5x = 30$$
$$x = 6.$$

(ii) Use a common denominator of x to give

$$\frac{1}{x} + \frac{x^2}{x} = 2, \text{ i.e. } \frac{1 + x^2}{x} = 2.$$

Hence

$$1 + x^2 = 2x$$
$$x^2 - 2x + 1 = 0$$

a quadratic which can be solved

$$(x - 1)(x - 1) = 0$$

i.e. $x = 1$.

S14

Solve the equation $6x - 8 > 3x + 9$ to give $3x > 17$, and hence $x > 5.66$, so the smallest integer (whole number) will be 6.

S15

$5^{-2} = \dfrac{1}{25} = 0.04; \quad 5^{1/2} = \sqrt{5} = 2.236;$

$5^0 = 1; \quad 125^{-1/3} = \dfrac{1}{5} = 0.2.$

Hence, the order is

0.04, 0.2, 1, 2.236,

i.e. 5^{-2}, $125^{-1/3}$, 5^0, $5^{1/2}$.

S16

(i) $f(3) = 2^2 = 4$; (ii) $f(0) = (-1)^2 = 1$;
(iii) $f(-3) = (-4)^2 = 16$.

S17

(i) $fg(10) = f\left(\dfrac{1}{10}\right) = \dfrac{5}{10} = \dfrac{1}{2};$

(ii) $gf(10) = g(50) = \dfrac{1}{50}.$

S18

(i) $f^{-1}(x) = \sqrt{(x + 1)}$; note that $f(3) = 8$, and $f^{-1}(8) = 3$.
(ii) $f^{-1}(3) = \sqrt{4} = 2$ and -2.

EXAM TYPE QUESTIONS

Q1

(i) Follow through this flowchart and write down whatever you are asked to do.

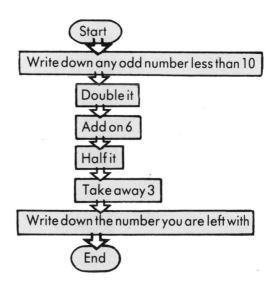

(ii) Find the number which is the same when it comes out as when it goes in through this flowchart.

Multiply by 2 → Take away 25 → Out

Q2

Use this flow diagram in each of the following questions

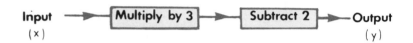

Input (x) → Multiply by 3 → Subtract 2 → Output (y)

(a) When the input is 4, calculate the output.
(b) When the input is 5.73, calculate the output.   (MEG)

Q3

Here are some 1 cm squares put together.

(a) Draw the next two patterns in the sequence.
(b) Complete the following table for this sequence.

Number of squares	1	2	3	4	5
Number of 1 cm edges	4	7			

(c) Describe the pattern of numbers in the bottom row of the table.
(d) How many 1 cm edges would be used in 100 squares?

Q4

Write down a possible value of x so that $2x < 4$ and $x > 1$.   (SEG)

Q5

Remember to use your answers in part (a), and to look for a pattern

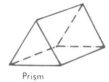
Prism

Cuboid

Pyramid

(a) Complete the table for the solids in Fig. 8Q.4.

Name	Number of vertices	Number of faces	Number of edges
Prism	6	5	9
Cuboid			
Pyramid			

(b) How many edges would you expect on a shape having 10 vertices and 7 faces?   (NEA)

Q6

Let f be the function $x \rightarrow \dfrac{4 - 3x}{2}$.

(a) Calculate f(−2).
(b) Find the value of x for which f(x) = −7.   (MEG)

Q7

The surface area, A, of a sphere with radius r is given by the formula $A = 4\pi r^2$.
(a) Calculate, to one decimal place, the surface area of a sphere with radius 4.5 cm.
(b) Rewrite the formula $A = 4\pi r^2$ to give r in terms of A and π.
(c) Calculate, to one significant figure, the radius of a sphere with a surface area of 580 cm².

Q8

(a) At the moment, Paul is six times as old as his daughter Jane.
    (i) If Jane's present age is x years, write down, in terms of x, the age Paul will be in 12 years' time.
    (ii) In 12 years' time, Paul will be three times as old as Jane. Write down an equation for x and hence find Jane's present age.
(b) Solve the equation $(x - 1)^2 = 3$, giving your answers correct to two decimal places. (MEG)

Q9

Which is the larger, 5^4 or 4^5?

Q10

(a) Find the value of k such that
    $(x - 3)(2x + 5) = 2x^2 + kx - 15$.
(b) Solve the quadratic equation
    $x^2 - 2x - 24 = 0$. (LEAG)

Q11

Write down the next two terms in the following pattern. Show how you found them.
1, 4, 12, 25, 43, 66, . . ., . . .

Q12

The stopping distance, d feet, at various speeds, v miles per hour, is given in the highway code as

Speed (mph)	30	50	70
Stopping distance (feet)	75	175	

but the stopping distance for 70 mph has been torn out. You are told that d and v are connected by the relation $d = av + bv^2$, where a and b are numerical constants.
(a) Use the table to find the values of a and b.
(b) Find the value of the stopping distance for 70 mph which has been torn out of the highway code table.

OUTLINE ANSWERS TO EXAM QUESTIONS

A1

(i) You should have written down the same odd number less than 10 twice.
(ii) If you try various numbers you should eventually find the answer of 25.

A2

(a) $(4 \times 3) - 2 = 10$; (b) $(5.73 \times 3) - 2 = 15.19$.

A3

(a) You should have drawn:

(b) The table will be:

Number of squares	1	2	3	4	5
Number of 1 km edges	4	7	10	13	16

(c) The numbers in the pattern go up by 3 each time from the 4, or it's the multiple of 3 with 1 added on each time.

(d) We can see that for n squares you would have $(1 + 3n)$ 1 cm edges, so for 100 squares you would need 301 edges.

A4

If $2x < 4$ then $x < 2$, hence $1 < x < 2$. So any value of x between 1 and 2 (but NOT equal to them) will do.

A5

(a) The completed table will be

	Vertices	Faces	Edges
Prism	6	5	9
Cuboid	8	6	12
Pyramid	4	4	6

(b) By looking at the numbers you will find that in each case the number of vertices, V, added to the number of faces, F, is 2 more than the number of edges, E. In other words, $V + F = E + 2$. So, with $V = 10$ and $F = 7$ we have the equation $10 + 7 = E + 2$, which will give us $E = 15$.

A6

(a) $(4 + 6)/2 = 5$.

(b) Solve $\dfrac{4 - 3x}{2} = -7$, hence $4 - 3x = -14$

$18 = 3x$ hence $x = 6$.

A7

(a) By substitution we get $A = 4 \times \pi \times 4.5^2$ which is 254.469, which rounds to 254.5.

(b) $A = 4\pi r^2$ will change to $\dfrac{A}{4\pi} = r^2$, which will then change to $r = \sqrt{\dfrac{A}{4\pi}}$.

(c) Substitute into the new formula to get $r = \sqrt{\dfrac{580}{4\pi}}$, which gives 6.79 to round off to 7.

A8

(a) (i) $6x + 12$; (ii) $6x + 12 = 3(x + 12)$

$$6x + 12 = 3x + 36$$
$$3x = 24$$
$$x = 8.$$

(b) $(x - 1) = + \sqrt{3}$ or $- \sqrt{3}$

$x = 1 + \sqrt{3}$ or $1 - \sqrt{3} = 1 + 1.73$ or $1 - 1.73 = 2.73$ or -0.73.

A9

5^4 is $5 \times 5 \times 5 \times 5$ which is 625, and 4^5 is $4 \times 4 \times 4 \times 4 \times 4$ which is 1024; so 4^5 is larger than 5^4.

A10

(a) Multiply out left-hand side to give $2x^2 - x - 15 = 2x^2 + kx - 15$; hence $k = -1$.

(b) $(x + 4)(x - 6) = 0$

$x = -4$ and $x = 6$.

A11

Next two terms are 94, 127. You should show the differences between each term, and that they go up in fives each time, i.e. Differences: 3 8 13 18 23

$$5 \ 5 \ 5 \ 5$$

A12

(a) By substituting the known values of $v = 30$, $d = 75$, and $v = 50$, $d = 175$ into the relation $d = av + bv^2$, we get the pair of simultaneous equations:

$$175 = 50a + 2500b \quad \ldots(1)$$
$$75 = 30a + \ 900b \quad \ldots(2)$$

If we multiply equation (1) by 3 and equation (2) by 5 we get:

$$525 = 150a + 7500b \quad \ldots(3)$$
$$375 = 150a + 4500b \quad \ldots(4)$$

We can now eliminate a by subtracting equation (4) from equation (3) to give:

$$150 = 3000b$$

which solves to $b = 0.05$. Now substitute this into the simplest equation (2) so that $75 = 30a + 45$, which solves to $30 = 30a$ or $a = 1$. So we have the answer $a = 1$, $b = 0.05$.

(b) Put the found values a and b into the relation $d = av + bv^2$ to give $d = v + 0.05 \times v^2$. Now substitute $v = 70$ into this to give $d = 70 + 0.05 \times 4900$ and you will see that $d = 315$. Therefore the stopping distance for 70 mph will be 315 feet.

G R A D E   C H E C K L I S T

FOR A GRADE F

You should now know:
> what a formula is;
> what a flowchart is;

and be able to:
> use letters to represent numbers, e.g. area of rectangle = $l \times b$;
> substitute numbers for words or letters in a simple formula.

FOR A GRADE C

You should also understand:
> the words (as used mathematically) expand, simplify, indices;

and be able to:
> use letters for numbers;
> generalise a simple number pattern;
> use variables in a computer program;
> find a formula to express an observed relationship;
> interpret and use brackets in a simple algebraic expression;
> factorise simple formulae;
> calculate any variable in a simple formula given the numerical values of the other variables;
> solve simple linear equations;
> use positive indices with letters and numbers.

FOR A GRADE A

You should also understand:

what is meant by a function;

and be able to:

add, subtract, multiply and divide algebraic fractions;

change the subject of a formula such as $s = 2\pi r(r + h)$ to h;

remove and insert brackets in algebraic expressions;

use positive, negative and fractional indices in both numerical and algebraic work;

factorise quadratic equations;

solve quadratic equations, simultaneous equations, fractional equations and inequations.

STUDENT'S ANSWER - EXAMINER'S COMMENTS

QUESTION

1	2	3	4	5	6	⑦	8	9
10	11	12	13	14	15	16	17	18
19	20	21	22	23	24	25	26	27
28	29	30	31	32	33	34	35	36
37	38	39	㊵	41	42	43	44	45
46	47	48	49	50	51	52	53	54
55	56	57	58	59	60	61	62	63

The diagram shows a number grid with a T drawn on it. It is said to be "centred at 7" because the branches of the T meet at 7.

(a) Find the total of the five numbers in the outline when it is "centred at 40"

$$39 + 40 + 41 + 49 + 52$$
$$= 221$$

A careless mistake. Should be 58.

(b) Complete the diagram below to show the five numbers in the T when it is "centred at x"

X–1	Ⓧ	X+1

X+9

X+18

good answer

(c) Show that the five numbers in a T "centred at x" will total 5x + 27

$$5x \; 44 + 27$$
$$= 247$$

| 43 | 44 | 45 |

53

62

$$= 247$$

Poor answer. Not shown from (b) that these expressions add up to 5x + 27. Only used one specific example.

(d) Find the five numbers in the T that totals 247.

43 44 45

53

62.

Found the correct ones but no indication of how.

(e) Explain why the total of the five numbers in the T could not be 240.

Not a good attempt, since the equation $5x + 27 = 247$ should have been solved

BECAUSE 3 CONSECUTIVE NUMBERS ADDED TO 2. NUMBE

9+18 HIGHER THAN THE MIDDLE NUMBER DO NO EQUAL

= 240

GRAPHS

The importance of a *graph* is that it can be used to interpret information, giving a visual picture of the information or data. That picture may, for example, indicate a *trend* in the data, from which we might be able to make future predictions. The graph might also be used to approximate a *solution* to a particular situation.

All the graphs used are on what we call *rectangular co-ordinates*; that is, the axes are at right angles to each other. When reading from graphs, or drawing on them, the kind of accuracy looked for in the examinations is usually no more than 1 mm out (some Examination Boards will insist the error must be *less* than 1 mm). So be as accurate as you can in both reading from graphs and in drawing on them.

USEFUL DEFINITIONS

Linear   a linear *equation* is one which involves no powers (other than 1 or 0), and no variables multiplied together, e.g. $x + y = 8$, or $3x = 4y - 2$
a linear *graph* will be a straight line

Quadratic   a quadratic *equation* is one which has a square as the highest power, e.g. $x^2 + 3x - y = 5$
a quadratic *curve* is a graph of a quadratic equation; it is a symmetrical **U** shape (or an upside-down **U**)

Cubic   a cubic equation is one which has a cube as the highest power, e.g. $y^3 - 6x^2 + 3xy = 0$

Gradient   the 'steepness' of a line, where the bigger the gradient the steeper the line 'uphill'; a negative gradient will be a line sloping 'downhill'

Intercept   where a line crosses an axis; an intercept on the *x*-axis is where a line crosses the *x*-axis; similarly for the *y*-axis

CO-ORDINATES
TRAVEL GRAPHS
DISTANCE/TIME GRAPHS
VELOCITY/TIME GRAPHS
SKETCH GRAPHS
INEQUALITIES
SIMULTANEOUS EQUATIONS

E S S E N T I A L   P R I N C I P L E S

1 > CO-ORDINATES

Co-ordinates are *pairs* of numbers that fix a particular position on a grid with reference to an origin.

The lines that go through the origin and have the numbers marked on them are called the *axes*; the horizontal one is called the *x*-axis, and the vertical one is called the *y*-axis. We always place the number representing the *horizontal* axis *before* that representing the *vertical* axis.

The origin in diagram (a) is the zero co-ordinate (0, 0). The co-ordinate of point *A* is (2, 1) because to get from the origin to this point *A* you would move 2 along the *horizontal* axis then 1 up the *vertical* axis. The other points you see have co-ordinates *B* (1, 3), *C* (4, 0), *D* (0, 1).

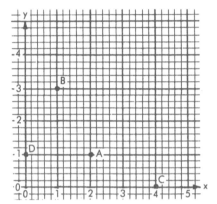

Of course, sometimes we wish to find the co-ordinates of a point that is not exactly on the lines labelled. For example, on diagram (b) the co-ordinates of *K* are $(1, 1\frac{1}{2})$ since *K* is 1 along the *x*-axis and $1\frac{1}{2}$ up the *y*-axis. In a similar way, the co-ordinates of *L* are $(1\frac{1}{2}, \frac{1}{2})$.

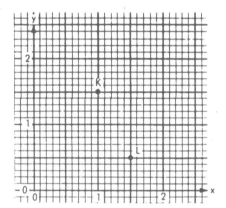

It is most important always to write and read co-ordinates in the correct way; the *first* number is *how many along*, the *second* number *how many up*. One way to remember this is **OUT** (Out and Up To it).

EXERCISE 1

Plot the following co-ordinates on a grid, with both axes going from 0 to 8, and join the points up in the order given:

(1, 4),   (3, 7),   (7, 5),   (5, 4),   (7, 3)   (1, 2)

DRAWING GRAPHS

One main use of co-ordinates is to help us draw graphs to assist us in sorting out information of one type or another. When we do this, the *x* and *y* axes are often labelled with other letters to help us see what the information is.

WORKED EXAMPLE 1

Using straight lines draw a graph from the given information about the costs of transporting weights by Blue Star Parcel Deliveries.

Weight (kg)	0	1	2	3	4	5
Cost (£)	1	2	3	6	7	8

(It is usual to draw graphs from tables like this with the top line giving you the horizontal axis.) We can choose a simple scale of 1 cm per kg along the horizontal and 1 cm per £1 up the vertical. Plot the co-ordinates from the table (0, 1), (1, 2), (2, 3) and so on to give the positions as shown. Then join up each point.

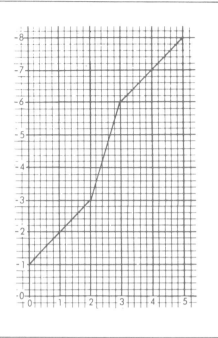

READING THE GRAPH

Emphasis in graphical work is on the extraction and interpretation of information displayed by graphs of various kinds. We shall now consider some of these with which you should be familiar.

CONVERSION GRAPHS

To help convert from one unit to another it is often helpful to have a handy chart or graph like this one, which shows the conversion of miles to kilometres.

You can find the number of kilometres approximately equal to any number of miles up to 30. Take, for example, 25 miles. From the 25-mile mark follow the vertical line up to the graph and you will see it is 40 kilometres, hence 25 miles is approximately 40 kilometres.

We can also work the other way round. For example, take 30 kilometres. From the 30-kilometre mark, follow the horizontal line along to the graph, and where you meet it come down the vertical line to 19 miles. Hence 30 kilometres is approximately 19 miles.

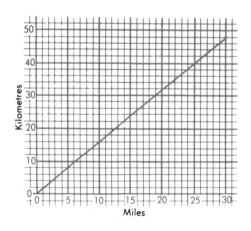

Notice too that each small line on the miles axis will represent a further mile since the 5-mile gap is divided into five equal parts, while on the kilometre axis each small line will represent 2 kilometres because the 10-kilometre gap is divided into five equal parts.

BRAKING DISTANCE

Another useful conversion graph is the braking distance graph. To help drivers realise that the faster they go the longer it takes to stop, the Ministry of Transport issued the following graph.

From the graph, you can see that each small horizontal line will represent 10 feet, while the vertical lines will represent 4 miles per hour. So, we can read from the graph that at 20 mph the stopping distance is approximately 30 feet, while at 70 mph (half way between the 60 and the 80) the stopping distance will be approximately 165 feet.

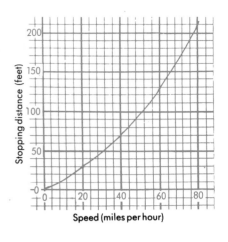

EXERCISE 2

The table shows the prices of different weights of new potatoes.

Weight (kg)	5	10	15	20	25
Cost (£)	1.20	2.40	3.60	4.80	6.00

(a) Plot the points on suitable axes to show this information on a graph.
(b) Use your graph to find (i) the cost of 18 kg of new potatoes, (ii) the weight of potatoes that can be bought for £2.

NEGATIVE CO-ORDINATES

You are supposed to be able to read and plot co-ordinates within the full range of negative and positive numbers. For example, on this grid you should see that the co-ordinates of the points are:

$$A(-2, -1), \quad B(-2, 0),$$
$$C(0, -1), \quad D(1, -2).$$

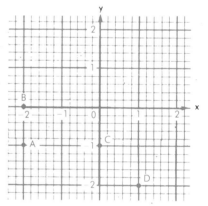

You should be able to recognise when to draw a straight line through points or draw a *smooth curve*. Generally, if the information you are plotting is only a small sample of a lot of possible data that the graph will eventually show (for example a time/distance graph), then, unless the points are in an obvious single straight line, we would always draw a smooth curve through them.

Draw a graph from the following information

Number of people	2	3	4	6
Hours to complete the job	6	4	3	2

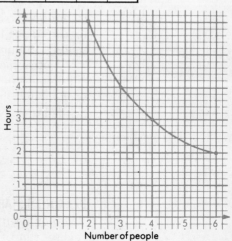

The points are not in a straight line, so we join up with a smooth curve.

NOTE

The straight line graphs are called *linear*, while the curved ones are *non-linear*.

CHOOSING SCALES

When you are faced with information that needs to be put onto a graph, you have to decide what scale to use. You are always going to get a more accurate graph the bigger it is drawn – but there is a limit to the size of paper available. You need to look at the largest numbers needed for each axis (and also if it needs to start at zero or not), then see how you can best fit this onto a scale that will fit the paper. Take care that you choose a scale where you can easily work out the position of in-between numbers. For example, a scale going up in 3's and having five divisions between each 3 is going to be a useless scale for reading in-between numbers. Your scale should ideally be going up in 1's, or 2's, or 5's, or 10's, ... When you've decided upon your scales you must fully label each axis of the graph with the necessary numbers on each darker line of the graph paper together with a description of what that axis is for. For example, velocity (mph) or time (seconds). Notice how the units are also written down (if there are any). Examples of this will be seen in the next section.

DRAWING GRAPHS FROM EQUATIONS

There are three main types of equation for which you should be able to draw graphs. (But check carefully those that are needed at each level.)

LINEAR EQUATIONS

A *linear* equation is of the form $y = mx + c$, where m and c are constants (a constant is a number that does not change at all, for example 3 or -5). This will always give a straight line. The minimum number of points to plot for a linear graph is three.

WORKED EXAMPLE 3

Draw the graph of $y = 4x - 3$ for x from -2 to 5.

We can see that the equation is linear, hence a straight line graph will be produced. We need at least three points to put in, so let's see what happens when $x = -2$, 0 and 5. A simple table of values can be made.

x	-2	0	5
$y = 4x - 3$	-11	-3	17

From this table you can see you will need 8 numbers on the x-axis (5 to -2), which will fit best with 1 unit to 2 cm. You will need 29 numbers on the y-axis (17 to -11), which will fit best with 1 unit to 1 cm. Draw these axes, label them, plot the three points $(-2, -11)$, $(0, -3)$ and $(5, 17)$, and join them up with a straight line.

QUADRATIC EQUATIONS

A *quadratic* equation is of the form $y = ax^2 + bx + c$, where a, b and c are constants. This will always give you a curved graph, and the part that is usually asked for is the part that does a U turn. See below.

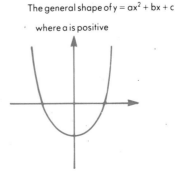

The general shape of $y = ax^2 + bx + c$ where a is positive

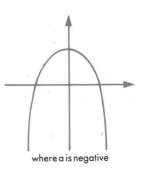

where a is negative

You need as many points as possible, especially round the 'dip'.

WORKED EXAMPLE 4

Draw the graph of $y = x^2 - 2x + 1$ from $x = -2$ to 3.

You can see from the equation that the general shape is a **U**-shaped curve, hence you will need a table of values with x from -2 to 3. It is usually best to build up the value of y as indicated in the table below.

x	-2	-1	0	1	2	3
x^2	4	1	0	1	4	9
$-2x$	4	2	0	-2	-4	-6
$+1$	1	1	1	1	1	1
$y = x^2 - 2x + 1$	9	4	1	0	1	4

From the table you can see that you will need 6 numbers on the x-axis (3 to -2), and this will fit best with 1 unit to 2 cm. You will need 10 numbers on the y-axis and no negative numbers, which will also fit best with 1 unit to 2 cm. Draw these axes, label them and plot the points $(-2, 9)$, $(-1, 4)$, $(0, 1)$, $(1, 0)$, $(2, 1)$ and $(3, 4)$, and join them up with a smooth curve.

EXERCISE 3

Draw the graph of $y = x^2 + 5x - 2$ from $x = -6$ to 2. Use the graph to solve the equation $x^2 + 5x - 2 = 0$.

RECIPROCAL EQUATIONS

A *reciprocal* equation is of the form $y = \dfrac{a}{x}$, where a is any integer and not equal to 0. This will give a curved graph and again you need as many points given as possible.

WORKED EXAMPLE 5

Draw the graph of $y = \dfrac{12}{x}$ from $x = -12$ to 12 ($x \neq 0$).

Although we want as many points as possible, 24 points is excessive. Since the x is to be divided into 12, we could just choose numbers that divide exactly into 12, i.e. the factors of 12, positive and negative. Remember we do not consider $x = 0$. So the table of values will be:

x	-12	-6	-4	-3	-2	-1	1	2	3	4	6	12
$y = \frac{12}{x}$	-1	-2	-3	-4	-6	-12	12	6	4	3	2	1

From the table you will see we need 25 numbers on the x-axis, which will fit best with five units to 2 cm. You will need 25 numbers on the y-axis also, and this will fit best with one unit to 1 cm. Draw these axes, plot the given points and join up with a smooth curve as shown in the figure.

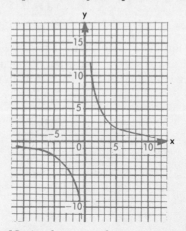

Notice how you have two separate curves; they do not join (although it is interesting to speculate what happens at $x = 0$ – study A level maths to find out!).

GRADIENTS

You will see on this distance/time graph that Paul drove from home 20 miles in the first hour when his average speed was 20 mph. In the next hour he drove a further 40 miles, hence his average speed was then 40 mph. During the last 2 hours he drove only 20 miles, which is 10 mph. We get an indication of the speed from how steep the lines are. The steeper the line, the greater the speed.

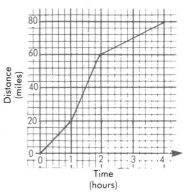

From the line, we can calculate accurately, the *gradient*. This is a measure of how steep the line is. We measure the gradient of a straight line by calculating between any two points on the line the difference in their vertical co-ordinates divided by the difference in their horizontal co-ordinates. Examples are given below.

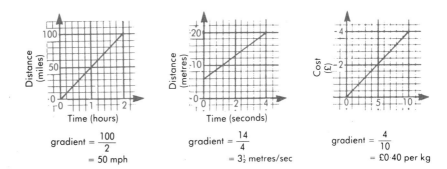

$$\text{gradient} = \frac{100}{2}$$
$$= 50 \text{ mph}$$

$$\text{gradient} = \frac{14}{4}$$
$$= 3\tfrac{1}{2} \text{ metres/sec}$$

$$\text{gradient} = \frac{4}{10}$$
$$= £0.40 \text{ per kg}$$

Notice, too, how the units of the gradient come from the labels on the axes. On each of these examples, the whole line was used to find the gradient. This is as accurate as possible, but you can take a shorter part of the line to find the gradient if necessary. Try out these examples for yourself, using only part of the lines and calculating (difference on vertical axis) ÷ (difference on horizontal axis).

NB: It is worth mentioning here that, for any linear equation of the form $y = mx + c$, the value of m will give the gradient of the line and c will say where the line crosses the y-axis (the *y-axis intercept*).

WORKED EXAMPLE 6

If you draw graphs from the following equations, all on the same axes, which will be the steepest line?

$$y = 3x + 7 \qquad y = x - 5 \qquad 3y = 6x + 7.$$

The gradient of $y = 3x + 7$ is 3, the gradient of $y = x - 5$ is 1, and to find the gradient of $3y = 6x + 7$ we need to divide throughout by 3 to get $y = 2x + 7/3$. This gives a gradient of 2. Hence the first equation of $y = 3x + 7$ gives the steepest line.

GRADIENTS ON CURVES

The sign of the value of m tells us whether the gradient is 'uphill' or 'downhill'.

If the m is positive then the graph will be 'uphill'.

If the m is negative then the graph will be 'downhill'.

But what about gradients on curves? Consider the following distance/time graph. This illustrates the speed of a ball thrown up to Michael at a window. He catches it and throws it back. While the ball is travelling up to Michael it will be slowing down, and this is illustrated on the curve between *A* and *B*. The curve starts steeply then slowly gets less steep. While Michael is holding the ball its speed is zero, hence the flat line *BC*. But when Michael throws the ball back its speed increases (or accelerates) as the graph shows the curve getting steeper (even if it is downhill).

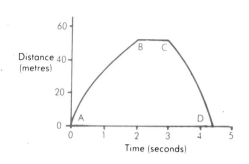

So, on distance/time graphs curves will be used to indicate gradual changes of speed as acceleration (getting quicker) or deceleration (getting slower). Below we shall look more closely at distance/time graphs.

> **2 ▷ TRAVEL GRAPHS**

❝ **You must become familiar with this type of question as it will very often occur.** ❞

Travel graphs are used to illustrate a journey of some sort. For example let's study this travel graph of Helen swimming.

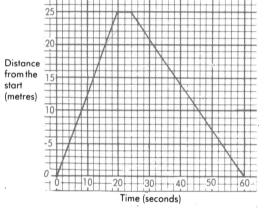

The graph shows Helen swimming the first length in 20 seconds, taking 4 seconds to turn around and set off back. Coming back she is much slower, probably because she is either tiring or doing a different stroke to the first length. We can work out how fast Helen swam the first length by seeing that 25 metres was swum in 20 seconds. This will give 75 metres per minute, which is $75 \times 60 = 4500$ metres per hour, or $4\frac{1}{2}$ kilometres per hour!

> **WORKED EXAMPLE 7**

The travel graph below shows James's journey from home to town. He walked from home to a bus stop, waited, then caught the bus to town.

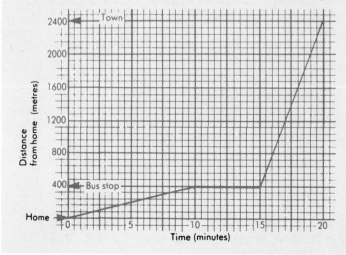

(a) How long did it take James to walk to the bus stop?
(b) If he left home at 1.55 p.m., at what time did he arrive in town?
(c) What is the distance between the bus stop and town?
(d) What was the average speed of the bus in kilometres per hour?

(a) The bus stop is 400 metres away, and the graph is at 400 after 10 minutes.
(b) It took James 20 minutes to get to town, and 20 minutes later than 5 minutes to 2 is 2.15 p.m.
(c) Town is 2400 metres away, the bus stop is 400 metres, and so the difference is 2000 metres.
(d) The bus covers 2000 metres in 5 minutes, that is (2000 × 12) metres in an hour, which is 24 000 metres. Hence the bus is travelling at an average speed of 24 kilometres per hour.

---

3 ▷ DISTANCE/TIME GRAPHS

This graph represents a gradual slowing down of, maybe, a cyclist. But how do we find the speed at any particular time?

To find the speed (velocity) at any given time we need to find the gradient at that point.

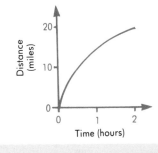

WORKED EXAMPLE 8

From the distance/time graph given below, find the velocity after 30 minutes.

> The tangent as you see is the line that just touches the curve and once only.

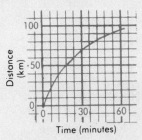

You need to draw the tangent to the curve where time is 30 minutes as shown:

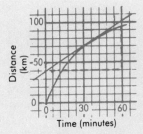

From the drawn tangent, calculate its gradient from the largest convenient triangle as shown. Here it is 60 km along the vertical divided by 1 hour along the horizontal, giving a gradient which is the velocity of 60 km/h.

EXERCISE 4

This graph shows the speeds of three different cars when being driven away from a standing start over a time of 25 seconds.
(a) From the graph, what is the speed of the Rover after 12 seconds?
(b) How long does it take each car to reach the speed of 40 km/h?

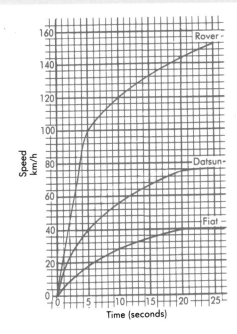

The graph shows a steady increase of speed from 0 to 40 mph over the first hour, then a steady 40 mph for the next hour. The increase in speed is the *acceleration*, and is measured by the gradient of the line.

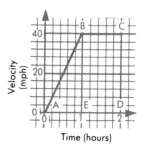

From this type of graph we can also find the distance travelled, by finding this area under the line. So, in this example, the distance travelled in the first hour is given by the area of the triangle *ABE*. The total distance travelled will be given by the area of the trapezium *ABCD*.

AREA UNDER CURVES

The area under curves on graphs can often give us additional information, as the following examples will illustrate.

This graph illustrates the flow of water through a pump. The area under the curve tells us exactly how much water was pumped in 3 seconds. The shaded area represents 20 litres of water. Hence in this case approximately 60 litres would have been pumped in the 3 seconds, as the whole area is approximately three times the shaded area.

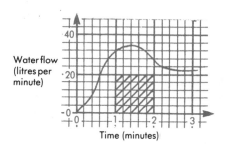

This graph illustrates a runner in a race. To tell how far he ran we need the area under the curve. The shaded area represents 25 metres, and the total area under the curve is approximately four of these squares. Hence the runner ran for 25 × 4, which is 100 metres.

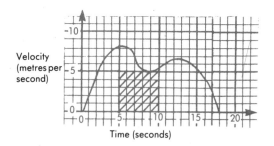

EXERCISE 5

From the velocity/time graph shown,
(a) calculate the speed after $1\frac{1}{2}$ minutes;
(b) calculate the total distance travelled;
(c) suggest what event is taking place.

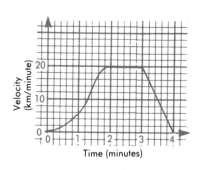

It is always helpful to be able to sketch the graph you are going to draw, but at the higher level of GCSE it is essential that you can sketch graphs reasonably accurately.

LINEAR EQUATIONS

You should already know that the shape of the graph from a linear equation $y = mx + c$ is a straight line with m the gradient and c the y-axis intercept (where the line crosses the y-axis). Hence you can sketch the graph from a linear equation quite easily.

WORKED EXAMPLE 9

Sketch the graph of $y + 2x = 10$.

Rewrite the equation to make y the subject, $y = 10 - 2x$ or $y = -2x + 10$, when we see that the gradient is -2, hence 'downhill', and that the graph cuts through the y-axis at $y = 10$.

So your sketch could look like the one shown.

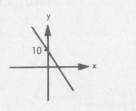

QUADRATIC EQUATIONS

If the equation is of a quadratic nature, i.e. $y = ax^2 + bx + c$, then it will be a **U**-shaped curve if a is positive, and a $\cap$-shaped curve if a is negative. The value of c is again the y-axis intercept. See what happens when $x = 1$. This should then give sufficient information for a rough sketch.

WORKED EXAMPLE 10

Make a sketch of the graph of $y = 3x^2 + 2x - 1$.

Since the 'a' is positive we know it's a **U** shape, since the 'c' is -1 we know it cuts through the y-axis at $y = -1$, and when $x = 1$, $y = 3 + 2 - 1$ which is 4. Hence a sketch could look like that shown.

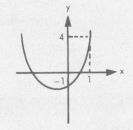

EXERCISE 6

Sketch the graphs of $x + y = 10$, $3x + 1 = y$ and hence obtain an estimate of their simultaneous solution.

When considering graphing an inequality such as $y < 3x$ we need to indicate all the points on the grid where this is true. You will soon see that you have a lot of points and *not* all in a straight line but all to one side of a straight line. The line in this case will be $y = 3x$. Now if we were graphing the inequality $y \leq 3x$ the solution would include the points on the line $y = 3x$ whereas the solution of $y < 3x$ will not include this line. So to sketch a region given by, say, $y > 4x$, we need first to draw the line $y = 4x$ then find which side of that line we want. One way to do this is to choose any convenient point that is *not* on the line itself and see if it is in the region or not. If it is, then shade in that region; if not, then shade the other region.

WORKED EXAMPLE 11

Shade on suitable axes the region $x + 2y > 6$.

Consider the line $x + 2y = 6$. It's a straight line and can be rewritten as $y = 3 - \frac{1}{2}x$. From this we can draw the line and then consider a point not on it. $(0, 0)$ is the easiest to consider, so substitute into $x + 2y > 6$ the values $x = 0$, $y = 0$, and see if it fits. However this gives $0 > 6$, which is false, so $(0, 0)$ is not in the region and we should shade the other side of the line, as shown in the diagram, to give us the region we require.

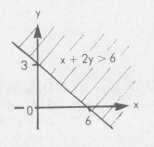

SOLUTION SETS

When we have to find a solution set from a number of inequalities, it is easier to shade *out* the regions we do not want, leaving unshaded the solution set.

 **WORKED EXAMPLE 12**

Illustrate the set of points that is satisfied by the inequalities: $y \geq 0$, $x \geq 0$, $x + y < 5$, $x + 2y > 6$.

$y \geq 0$ and $x \geq 0$ indicate to us that we only need the region where both x and y are positive. Thus, shading out the other two regions, as shown, gives us the points we are looking for.

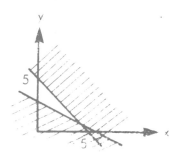

NOTE

You always need to be aware of the inequality sign and to notice whether it is $>$ or $\geq$. In the latter case you *do* need to include the points on the line in your solution sets.

EXERCISE 7

When taking bookings for a tour of, at the most, 14 people, a tour operator insists on taking at least twice as many women as men. Illustrate on a graph the solution set of possible combinations of men and women.

7 ▷ SIMULTANEOUS EQUATIONS

One way to solve these is to draw graphs from both equations. The point of intersection of the lines is the solution.

 **WORKED EXAMPLE 13**

❝ But only solve this way if you're told to in the question, because the method is not as accurate as algebra. ❞

Find the points that satisfy both equations $y = x^2 - x - 2$ and $y = 2x$.

Draw the graph of both equations on the same axes as shown. The graphs intersect at the points $(-\frac{1}{2}, -1)$, and $(3\frac{1}{2}, 7)$. Hence the solutions are $x = -\frac{1}{2}$, $y = -1$ and $x = 3\frac{1}{2}$, $y = 7$.

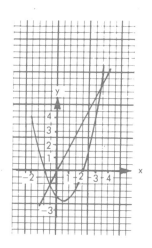

EXERCISE 8

The product of two numbers less than 8 is 12. One of the numbers is 1 less than the square of the other. By drawing suitable graphs, find the two numbers.

SOLUTIONS TO EXERCISES

S1

You should obtain an answer looking like this:

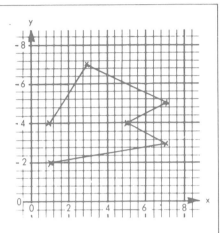

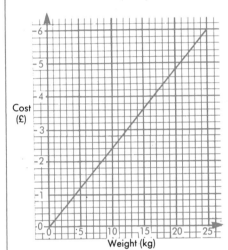

S2

(a) Your result should be like the graph shown.

(b) Read from *your* graph, but if your graph is like the one shown, then you should have answers very close to (i) £4.30 and (ii) 8.5 kg.

S3

You should have a table of values as below:

x	−6	−5	−4	−3	−2	−1	0	1	2
x^2	36	25	16	9	4	1	0	1	4
$5x$	−30	−25	−20	−15	−10	−5	0	5	10
−2	− 2	− 2	− 2	− 2	− 2	−2	−2	−2	− 2
y	4	− 2	− 6	− 8	−8	−6	−2	4	12

which gives a **U** shape.

The solution to $x^2 + 5x - 2 = 0$ is the points on the graph where $y = 0$, these points are where the graph cuts the x-axis and you should have answers of approximately −5.4 and 0.4.

S4

(a) Read from 12 on the horizontal axis up to the Rover graph, where taking a straight line horizontally to the speed axis it reads just above the 128 km/h, which will be 129 km/h.

(b) Reading along the 40 km/h line we initially come to the Rover line which is above approximately 1.3 seconds, then comes the Datsun line which is 5.5 seconds, and finally the Fiat which first reaches 40 km/h after 19 seconds.

S5

(a) By drawing a tangent to the curve at time equal $1\frac{1}{2}$ minutes you obtain a straight line with gradient of $25 \div 1.6$, which rounds off to 16 km/minute.

(b) The total distance travelled will be given by the area under the whole graph. This can be estimated at just under five squares, each representing 10 kilometres, hence the total distance will be approximately 50 km.

(c) The maximum speed is 20 km/h per minute which is 1200 km/h. This is very close to the speed of sound or very fast aircraft. The event could possibly be a short flight by a jet plane.

S6

They are both linear, hence two straight lines are wanted. A simple sketch should reveal the intersection to give $x = 2$, $y = 8$.

S7

If we use m for the men, and w for the women, then we can obtain two inequations, $m + w \leqslant 14$ and $w \geqslant 2m$. If we draw graphs of these two and shade out the regions we do not want, we obtain the diagram shown. The unshaded part represents all the possible combinations of men and women.

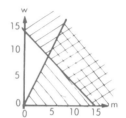

S8

Let the two numbers be x and y, and we then have two equations: $xy = 12$ and $y = x^2 - 1$. Neither is linear, so both give curves. Remembering that each number is less than 8, the possible tables of values of each equation could be:

x	6	4	3	2
$y = 12/x$	2	3	4	6

x	3	2	1	0
x^2	9	4	1	0
$y = x^2 - 1$	8	3	0	−1

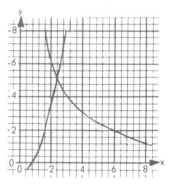

These you can plot on a graph. Join up both lines with smooth curves, and you should get an answer like the graph shown, which gives a solution approximately equal to $x = 2.4$, $y = 5$. Hence the two numbers are approximately 2.4 and 5.

EXAM TYPE QUESTIONS

Q1

The graph shows the distance of a train from London in km. Find the speed of the train in km/h at (a) 9.05 a.m., (b) 9.15 a.m.   (MEG)

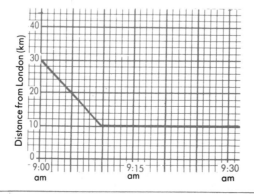

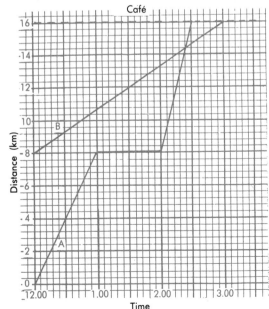

Q2

This conversion graph has been drawn to show the rate of exchange from English £ to German Marks in 1985.

(a) By drawing suitable lines on your graph paper, use the graph to find:
    (i) the number of German Marks equivalent to £10;
    (ii) the cost in £ of a watch bought in Germany for 50 Marks.

(b) A new exchange rate gives £1 = 3.75 Marks. On the graph draw a *new line* to represent this.

Q3

The travel graph shows the journey of two men, Albert (A) and Brian (B) who both set off at 1200 noon one day to meet at a café. Brian starts 8 km nearer the café than Albert, and walks steadily for 3 hours with no rests. Albert runs for 1 hour, then rests for an hour, before running to the café.

(a) What was (i) Albert's speed before his rest? (ii) Brian's speed?

(b) At what time did Albert overtake Brian?

(c) How long did Albert have to wait at the café for Brian to arrive?   (MEG)

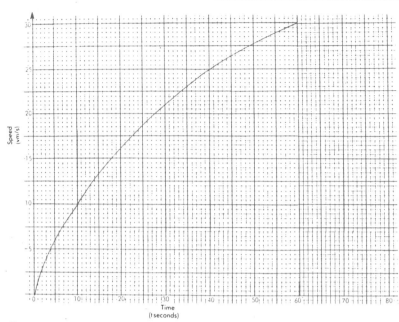

Q4

The speed of a car is observed at regular intervals of time. The velocity/time graph shown above has been derived from these observations.

(i) Use the graph to estimate the car's acceleration at $t = 20$.

(ii) State briefly how the acceleration of the car changes over the 60 seconds for which the graph is drawn.

(iii) Use the graph to estimate how far the car travels in the first minute.

(iv) At $t = 60$ the driver applies the brakes to produce a constant retardation of 2 m/s^2. Extend the graph to show this retardation and state the value of t when the car comes to a stop.

(NEA)

Q5

A firm calculated the pay, £P, of its employees who had worked for a number of years, Y, using the formula $P = 50 + 5Y$. It was suggested that a simple graph could be made to illustrate this information readily. Draw the graph from this formula, bearing in mind that it is unusual for anyone to work with this firm for more than 40 years.

Q6

A flagpole on the left casts a shadow on the ground. The length of the shadow alters with the time of day. The following graph shows the length of the shadow from 0900 to 1400 hours.

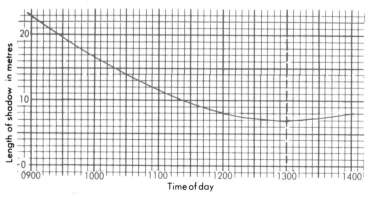

(a) How long was the shadow at 0930?

(b) At what time was the shadow 16 m long?

(c) The graph is symmetrical about the dotted line. Estimate the length of the shadow at 1500.

(d) When was the sun at its highest point in the sky?

(MEG)

Q7

The diagrams show vertical cross-sections of two cylindrical containers, A and B, and of two containers, C and D, each of which is part of a cone.

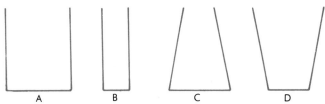

Each container is filled from a tap from which water is flowing at a constant rate. The graphs below show the depth of water measured against time in each of three of the containers.

Identify the container to which each graph refers.   (NEA)

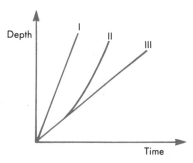

Q8

The distance/time graph illustrates two cyclists, Vijay and Neil, in a race. Describe what happened in the race.

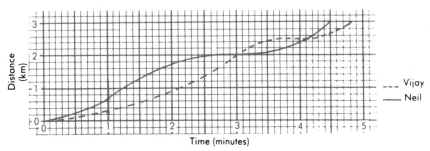

Q9

(a) Given that $y = 4x^2 - x^3$, copy and complete the following table:

x	0	0.5	1	1.5	2	2.5	3	3.5	4
y	0		3		8	9.375		6.125	0

Using a scale of 4 cm to represent 1 unit on the x-axis and 2 cm to represent 1 unit on the y-axis, draw the graph of $y = 4x^2 - x^3$ for values of x from 0 to 4 inclusive.

(b) By drawing appropriate straight lines on your graph
   (i) estimate the gradient of the curve $y = 4x^2 - x^3$ at the point $(3.5, 6.125)$,
   (ii) find two solutions of the equation $4x^2 - x^3 = x + 2$.   (MEG)

Q10

A farmer has a hopper with a square top (side length l). The volume of grain the hopper can hold is given, approximately, by the formula $V = 2l^2$.

(a) Draw a graph of how the volume changes as the side of the square hopper top, l, changes from 1 metre to 4 metres.

(b) What value of l will give a volume of 20 m³?

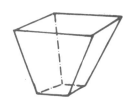

Q11

The table shows the income received by a small business as they increased their monthly expenditure on advertising.

Month	Jan	Feb	March	April	May	June	July
Expenditure in £'s	200	400	600	800	1000	1200	1400
Income in £'s	560	900	1120	1260	1360	1440	1480

(i) Plot this information on graph paper and draw a curve of best fit for these plots.

By drawing appropriate straight lines on your graph, find

(ii) how the rate of increase of income from advertising when the expenditure was £200 differed from that when the expenditure was £1200,

(iii) the expenditure when the income increased by £1 for every extra £1 spent.

OUTLINE ANSWERS TO EXAM QUESTIONS

A1

(a) Speed is gradient of the sloping line, which is 20 km in 10 min, hence 120 km/h.
(b) At 9.15 the train is not moving, hence speed is 0 km/h.

A2

(a) (i) £10 = 40 Marks
   (ii) 50 Marks = £12.50
(b) Line joined through (0, 0), (10, 37.5), or equivalent.

A3

(a) (i) Albert covered 8 km in 1 hour, hence his speed was 8 km/h; (ii) Brian travelled 8 km in 3 hours, an average of $8 \div 3 = 2\frac{2}{3}$ km/h. It is likely that an answer in between $2\frac{1}{2}$ and 3 km per hour (inclusive) would be acceptable.
(b) This happens where the two lines cross. Since there are 10 divisions represented by the lines between each hour, each one will represent 6 minutes. The lines cross over at 4 small lines past 2.00, which will be 24 minutes past 2. It would be quite acceptable to give any answer between (and including) 21 minutes and 27 minutes past 2.
(c) When the lines reach the 16 km line, that is the café. So the time waited is the difference of the time shown by the two ends on the café line; this is 30 minutes.

A4

(i)   Tangent at $t = 20$ has gradient of approximately 0.6 m/s²
(ii)   The acceleration is getting smaller.
(iii) Area under the curve, approximately 1000 m (1 km).
(iv)   Should be a straight line from (60, 30) to (75, 0), i.e. $t = 75$.

A5

This equation is of the form $y = mx + c$ and hence a straight line graph will be produced. The values of Y to look at will vary from 0 to 40, hence we need $Y = 0$, 20 and 40 to produce the table of values.

Y	0	20	40
$P = 50 + 5Y$	50	150	250

This indicates that the horizontal axis (Y) will need to show numbers up to 40, hence five units to 2 cm will be the best fit. The vertical axis (P) needs to represent numbers up to 250, which can be done with a scale of 10 units to 1 cm.

A6

(a) 19 m;   (b) 10.00;   (c) same as 11.00, which is 11 m; (d) when the shadow was shortest, that is 13.00.

A7

Graph I fills the quickest and at a steady rate, hence is container B.
Graph II gradually fills more quickly hence is smaller at top and so is container C.
Graph III fills steadily but the slowest, hence is container A.

A8

Your description should be written in sentences ('good English') and not in note form. It should include the following details: Neil takes the lead then gradually slows down to a standstill. Vijay overtakes Neil after 3 minutes but then slows down himself and Neil overtakes him. Neil then goes on to win the race after $4\frac{1}{2}$ minutes.

A9

(a) The missing values in the table are 0.875, 5.625 and 9.
(b) (i) By drawing a tangent at $x = 3.5$, gradient is about -8.75;
    (ii) By drawing the line $y = x + 2$ and finding the points of intersection, you find the solutions to be $x = -0.6$, 1 or 3.6.

A10

(a) Using the formula $V = 2l^2$ from $l = 1$ to 4 will give us the table:

l	1	2	3	4
l^2	1	4	9	16
$V = 2l^2$	2	8	18	32

The horizontal l-axis needs only 4 numbers; that will need a scale of 1 unit to 4 cm. The vertical V-axis needs 32 numbers, so use a scale of 5 units to 2 cm. Then plot the points from the table and draw a smooth curve.
(b) Reading from your graph when $V = 20$ should give you the approximate value of $l = 3.2$ metres.

A11

(ii) Tangents should be drawn for expenditure at 200 and at 1200, giving:
      at 200 the rate of increase (gradient) of 2
      at 1200 the rate of increase (gradient) of 0.6
(iii) You need the point that has a gradient of 1, this is where the Expenditure is £580.

G R A D E   C H E C K L I S T

FOR A GRADE F

You should now know:
    how to read graphs;
and be able to:
    read and plot points whose co-ordinates are positive;
    construct a graph from information given;
    read related pairs of values from a graph with straightforward scales on the
        axes;
    describe the main features of a graph given to you.

FOR A GRADE C

You should also be able to:

use co-ordinates to read or plot points for any value of x and y;

draw linear and non-linear graphs from given data;

determine appropriate scales for graphs, then draw and label axes;

draw graphs of the type $y = mx + c$, $y = ax^2$ and $y = \frac{a}{x}$, where a and b are integers;

interpret the meaning of m and c in the equation $y = mx + c$;

calculate and interpret the gradient of a straight line graph;

recognise the significance of the changing gradients of a non-linear graph.

FOR A GRADE A

You should also be able to:

sketch graphs of the functions $ax + b$ and $ax^2 + b + c$, where a, b, c are integers;

graph inequalities and simultaneous equations to give suitable solution sets;

make deductions and calculations from distance/time graphs and velocity/time graphs;

calculate the gradient of a curve at a point and interpret the result.

STUDENT'S ANSWER - EXAMINER'S COMMENTS

QUESTION

The graph given below is that of the function f which is defined by $f(x) = 10 - \dfrac{5}{x}$

(i) From the graph, find the value of x for which $f(x) = 4$.

(ii) g is the function $g(x) = 2x$.

   Draw the graph of the function g for $0 \leqslant x \leqslant 5$.

(iii) Use your graphs to solve the equation

$$2x = 10 - \frac{5}{x}.$$

   Give your solutions correct to one decimal place.

graph slightly out here, but good enough to get full marks

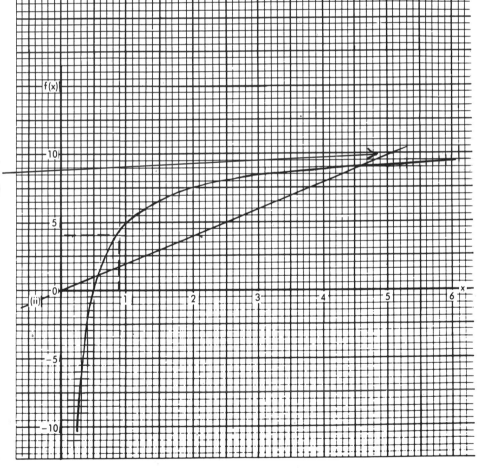

good - correct answers

Answers: (i) ___0·9___

(iii) $x = 0·6$ AND $4·6$

G E T T I N G   S T A R T E D

Geometry can be defined as the science of properties and relations of magnitudes (as lines, surfaces, solids) in space. The emphasis in the examinations will be on the well-established geometrical properties and relationships, and on how these can be used to convey information and to solve problems. As a result, this topic often appears as an important part of questions on *drawings*, *bearings* and even *algebra*, as well as in questions devoted *solely to geometry*. Many facts that you should learn and be familiar with are given in this chapter, so that when they arise within questions you can recognise the situation and apply the facts with confidence.

USEFUL DEFINITIONS

Angle	the amount of turn, measured in degrees
Transversal	a line that crosses through at least two parallel lines
Diagonal	a line joining two corners of a geometric shape
Subset	a part of a larger set
Circumference	the perimeter of a circle
Semi-circle	half of a circle
Vertex	a point where two lines, or edges, meet
Included angle	the angle in between two lines of defined length
Edge	the line where two faces meet
Face	surface of a solid shape bounded by edges
Cross section	the plane shape revealed by cutting a solid shape at right angles to its length (or height)

ANGLES
PARALLEL
PLANE FIGURES
POLYGONS
CIRCLES
SOLID FIGURES
SIMILARITY
CONGRUENCY
SYMMETRY

E S S E N T I A L   P R I N C I P L E S

1 > ANGLES

Every angle can be described by its size, and depending on this it falls into one of four main categories:

Acute angles are angles less than 90°
Right angles are angles that equal 90°
Obtuse angles are angles that are bigger than 90° but less than 180°
Reflex angles are angles that are bigger than 180° but less than 360°

There are situations that you should be familiar with and they are illustrated here:

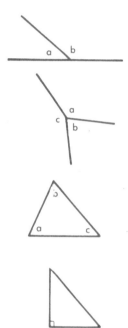

Angles on a line, as a and b shown here, will always add up to 180°.

Angles around a point, as a, b and c shown here, will always add up to 360°.

The three angles inside a triangle, as a, b and c shown here, will always add up to 180°.

A right angle is usually written as a box in the angle, as shown here. Any two lines that are at right angles to each other are said to be *perpendicular*.

2 > PARALLEL

Two lines are said to be parallel if the perpendicular distance between them is always the same.

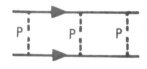

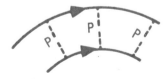

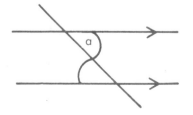

As you will see, parallel lines are not necessarily straight, but in most situations if you are told two lines are parallel you should assume that they are straight lines unless you have a very good reason not to.

The diagram shows a pair of parallel lines and a *transversal* cutting them. The angles marked a and b, which are called *alternate angles*, are always equal to each other.

3 > PLANE FIGURES

You should recognise, be able to name and know the following facts about each shape below:

An **isosceles triangle** has two of its sides the same and two angles the same, as indicated on this diagram. Sides of the same length are marked.

Equal angles are marked with ⊾.

An **equilateral triangle** has all its three sides the same length and all its angles are 60°.

A **right-angled triangle** is one that contains a right angle.

A **quadrilateral** has four sides, and the four angles it contains will add up to 360°.

A **rectangle** has four sides and its opposite sides equal, as shown, and all its angles are right angles.

A **square** has all its four sides equal and all its angles are right angles.

A **kite**, recognisable as a kite shape, has four sides, as shown, the top two sides with the same length and the bottom two sides with the same length.

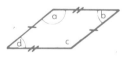

A **parallelogram** has four sides and the opposite sides are of equal length as shown. The opposite sides are parallel. In a parallelogram the angles next to each other will always all add up to 180°. For example, $a + b = b + c = c + d = d + a = 180°$. Also, the angles opposite each other will be equal. For example, $a = c$, $b = d$.

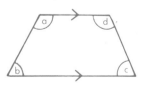

A **trapezium** is a quadrilateral that has two parallel sides, the pairs of angles between each parallel side add up to 180°. For example, in the trapezium drawn here $(a + b) = 180°$ and $(d + c) = 180°$.

A **rhombus** is a parallelogram that has all its sides the same length. The *diagonals* of a rhombus are *perpendicular*. This last fact will often be needed in examinations and should be learnt.

this one fact is so often forgotten or ignored in exams. Do learn it. 💬

SETS

It is useful to remember that:

{square} ⊂ {rhombus} ⊂ {kite} ⊂ {quadrilateral}

{square} ⊂ {rectangle} ⊂ {parallelogram} ⊂ {trapezium} ⊂ {quadrilateral}.

It is a useful exercise to put the above sets into a Venn diagram.

4 ▷   POLYGONS

You ought to be familiar with the names of the polygons mentioned below. **Polygons** are 'many-sided two-dimensional shapes'.

Triangle – 3 sides	**Quadrilateral** – 4 sides	**Pentagon** – 5 sides
Hexagon – 6 sides	**Septagon** – 7 sides	**Octagon** – 8 sides
Nonagon – 9 sides	**Decagon** – 10 sides	

Polygons have two main types of angles. There are *interior* angles (inside) and *exterior* angles (outside), as illustrated.

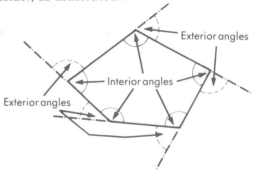

A polygon will have as many interior as exterior angles, which will be the same as the number of sides of the polygon. You should know the following facts about these angles:

Exterior: all the exterior angles of any polygon always add up to 360°.

Interior: all the interior angles of any *N*-sided polygon always add up to $180(N - 2)°$.

Regular polygons are polygons with all the sides the same length and all interior angles the same (hence all exterior angles will be the same also). We can use these last two facts to find their angles.

WORKED EXAMPLE 1

Calculate the size of the exterior and interior angles of a decagon.

There will be 10 equal exterior angles, all adding up to 360°, hence the size of each one will be 360 ÷ 10 which is 36°. You should be able to see from the diagram above that the exterior angle and the interior angle add up to 180°. Hence, if we've just calculated the exterior angle to be 36°, then the interior angle will be 180° − 36°, which is 144°.

EXERCISE 1

Calculate the interior angle of an octagon.

5 ▷   CIRCLES

Any straight line drawn from the centre of a circle to the edge of that circle (*circumference*) is called a *radius*. In any circle you can draw (if you wish to) hundreds of radii (plural of radius) all of which would be the same length. A straight line drawn from one side of a circle to the other side, passing through the centre, is called a *diameter*. Again, any circle will have hundreds of diameters all of the same length.

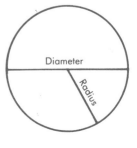

Any straight line drawn in a circle from one part of the circumference to another, as illustrated, is called a *chord*. The two parts of this circle have been split into *segments*; the smaller one is called the minor segment and the larger one is called the major segment.

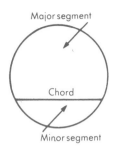

There are a number of interesting facts about angles in a circle that you should know. You can test all of them for yourself by drawing your own examples.

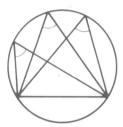

From any chord in a circle there are many triangles that can be formed in the same sector that touch the circumference, as illustrated. All these angles opposite the chord will be *equal*.

From any chord in a circle there is only one triangle that can be drawn to the centre of the circle, and this angle will be *double* any angle drawn to the circumference in the same sector as the centre. For example, in the diagram the angle at the centre is $2x°$ and at the circumference it will be $x°$.

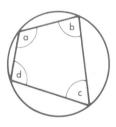

Any quadrilateral drawn so that its four vertices touch the circumference of a circle is said to be *cyclic*. Its opposite angles will add up to 180°. For example, in the diagram shown $(a + c) = (b + d) = 180°$. It is also true that any quadrilateral that has opposite angles adding up to 180° is cyclic and hence a circle can be drawn around the vertices.

SEMI-CIRCLE

If you draw any triangle in a semi-circle where one side is the diameter, as illustrated, then the angle made at the circumference will always be a *right* angle.

TANGENTS

A tangent to a curve, or a circle, is a line that will touch the curve or circle at only one point. If drawn on a circle this tangent will be perpendicular to a radius.

There are therefore two ways to draw a tangent on a circle at a particular point. One is put your ruler on that point and simply draw the line that only touches the circle there. The other way is to construct a right angle at that point on the radius and hence draw in the tangent. (See page 164.)

EXERCISE 2

D and C are the centres of the two circles. A and B are the points where the common tangent touch the circles.

(i) What is the full name of the quadrilateral $ABCD$?

(ii) When will $ABCD$ be cyclic?

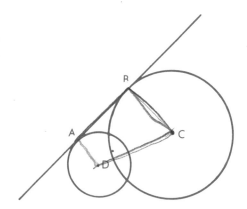

6 ▷ **SOLID FIGURES**

You should be able to recognise and name the following *solid shapes* and be able to construct some of them from suitable material like card or straws.

A **cube** has all its sides the same length.

A **rectangular block** (or *cuboid*) has each opposite side the same length.

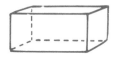

A **sphere** is just like a football or a tennis ball.

A **cylinder** is like a cocoa tin or a drain pipe, with circular ends.

A **cone** is like an upside down Cornetto, with a circle for the base and a smooth curved surface rising to a point at the top (like a witch's hat!)

A **pyramid** can have any shape for its base, but then from each side of the base the sides of the pyramid will meet at a point as shown. This diagram would be called a 'square based pyramid' since the base is a square. When the vertex is perpendicularly above the centre of the base the correct name is a 'right pyramid'.

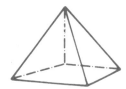

PRISMS

Any three-dimensional shape with the *same cross section through its length (or height)* is called a **prism**.

For example, consider the shapes below. They are all prisms, since they are shapes you could 'slice' up in such a way that each cross section would be identical.

You should be familiar with the words used to describe the different features of solids (see Q5 in Chapter 8). These are:

Face   the flat surfaces of solid shapes
Edge   lines where faces join together
Vertex a point where edges join.

VIEWS

When looking at solid shapes you get different views from different positions. Two important views are *plans* and *elevations*.

PLANS

The *plan* of a shape is the view you get when looking down from directly above the shape.

ELEVATION

The *end elevation* of a shape is the view you get when looking at the *end* of the shape. The *front elevation* of a shape is the view you get when you look directly at the *front* of the shape.

WORKED EXAMPLE 2

This diagram represents one of the great Pyramids. Draw a sketch of the view you would get of this pyramid when
(a) looking directly from one side (end elevation);
(b) looking down from above (plan).

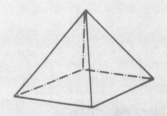

The end elevation will be

and the plan will be

NETS

A **net** is a flat shape that can be folded up to create a solid shape.

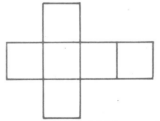

This net would fold up to make a *cube*.

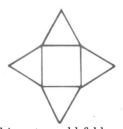

This net would fold up to make a *square based pyramid*

You need to be able to recognise what shape a net will fold into, and also to draw a net for yourself for any given shape. There is usually more than one possible net that will give any shape. For example, try to find at least two more nets that will make up the cube. But, do beware: when actually making a real net you would put 'tabs' onto a number of sides so that you could glue the shape together, whereas an examination question (as the diagrams here) will not usually have them shown or expect you to put them onto the shape when drawing them.

EXERCISE 3

The figure shows the net of an open box. The box will be 20 cm long, 10 cm wide and 5 cm high.

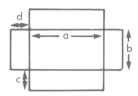

(i) If this net was the right size for the box described, what would be the lengths of a, b, c and d?

(ii) What are the dimensions of the smallest rectangular piece of card that could be used to cut out this net?

7 > SIMILARITY

Any two shapes are said to be *similar* if all the angles that could be drawn and measured in the shapes are the same, and if one shape is the same as the other but a different size. (See the reference to Similarity in Chapter 7.)

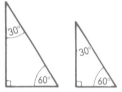

These two triangles are similar because all their angles are the same. It is true to say that if any two triangles have exactly the same angles as each other they will be similar.

As well as looking at the angles in other shapes we shall need to see if one shape can be enlarged to be the same as the other. You will meet this idea later in transformation geometry.

WORKED EXAMPLE 3

A builder built a garage with a floor similar to this rectangle. The garage was built with a width of 2 metres. How long was it?

Since the two shapes are similar, then the enlargements of the sides will be the same. The width has gone from 1 cm (if you measure it) to 2 metres. That is 200 times larger, so the length will go from 3 cm (if you measure it) to 3×200, which is 600 cm. So the garage was 6 metres long.

8 > CONGRUENCY

Two shapes are *congruent* if they are exactly the same shape and size. This means also that the angles of one shape would be the same as the angles of the other shape.

You will be expected to recognise when two triangles are congruent. The following examples illustrate the minimum information needed to determine whether two triangles are congruent or not. All three sides are identical, so we say *ABC* is congruent to *YXZ*. (Note that the letters should *correspond* with the same angles, i.e. angle at A is the same as the angle at Y, B is equal to X.)

All three angles and a corresponding side are equal, so we say *ABC* is congruent to *NML*. (*Note*: As seen above, if only two angles are given then the third is always known.)

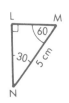

Two sides and the angle in between (included) are equal, so we say *ABC* is congruent to *PQR*.

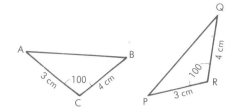

WORKED EXAMPLE 4

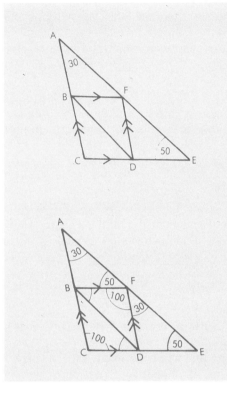

From this diagram name two congruent triangles.

From what we have learnt about angles around a parallelogram we can fill in some of the other angles, as shown on this diagram. The line from *B* to *D* is a transversal of the parallelogram *BFDC*, hence we have two alternate equal angles marked. We can now see that △*BFD* and △*DCB* have the same angles of 100°, ∡ and the third angle (which is 180 − (100 + ∡), but we do not need to find this out). They also have the line *BD* in common, hence we have here the situation where we have 'all three angles and a corresponding side being equal'. Therefore △*BFD* is congruent to △*DCB*.

EXERCISE 4
State two triangles in the diagram that are:
(a)   congruent;
(b)   similar but not congruent.

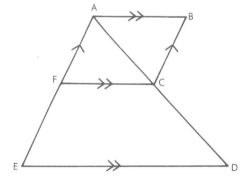

NOTE
It is worth noticing that shapes that are congruent are obviously *similar*, but similar shapes are only congruent when *identical*.

9 > SYMMETRY

There are two particular types of symmetry to be familiar with: *line* symmetry and *rotational* symmetry.

LINE SYMMETRY
If you can fold a shape over so that one half fits exactly on top of the other half, then the line over which you have folded is called a *line of symmetry*. The following examples will illustrate this. The dotted lines are the lines of symmetry.

The square has four lines of symmetry, the rectangle two, the isosceles triangle just one, as has the quadrilateral next to it, while the circle has thousands and thousands of lines of symmetry: there are too many for us to count (we call this an *infinite* number).

Often, of course, you cannot fold over a shape that you are looking at, so you either have to imagine it being folded or trace it on tracing paper and then fold it. In most examinations you would be allowed to trace the shape and fold it over to find lines of symmetry.

EXERCISE 5
(i)   Sketch a hexagon with only one line of symmetry.
(ii)   What other possible number of lines of symmetry can a hexagon be drawn with?

ROTATIONAL SYMMETRY
This is sometimes also called 'point symmetry'. A square has *rotational* symmetry of order 4, because if you turn it round its centre there are four different positions that it can take that all look the same, as shown in this illustration.

Try this out with a rectangle (use this book), and you should find the rectangle has rotational symmetry of order 2.

Any shape that has what we would call 'no rotational symmetry', such as an elephant shape or an 'L' shape, has rotational symmetry of order 1, since there is only one position where it looks the same! By the way, a circle will have rotational symmetry of an infinite order.

EXERCISE 6
(a)   In each letter below, draw (if any) all the lines of symmetry.

MATHS

(b)   Which of the above letters have rotational symmetry of order 2?

POLYGON SYMMETRY
All regular polygons will have the same number of lines of symmetry as the order of rotational symmetry, which is the same as the number of sides. Look at a square or at any regular polygon and check this out for yourself.

3D SYMMETRY
The symmetry of 3D shapes is of two types, as in 2D shapes.

PLANES OF SYMMETRY
These are similar to lines of symmetry in 2D. A shape has a *plane of symmetry* if you can 'slice' the shape into two matching pieces one the exact mirror image of the other. To find these planes of symmetry you need to be able to visualise the shape being cut and to see in your mind whether the pieces are matching mirror images or not.

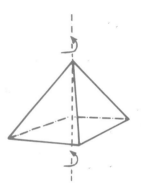

WORKED EXAMPLE 5

Find how many planes of symmetry a cuboid has.

Consider the cuboid shown above. We can cut it into two exact halves in the following three ways. Hence the shape has three planes of symmetry.

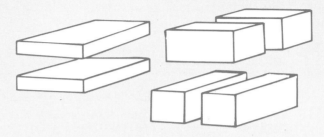

EXERCISE 7

Find how many planes of symmetry a rectangular based right pyramid has.

AXES OF SYMMETRY

An *axis of symmetry* is a line through which the shape may rotate and yet still occupy the same space. For example, in the square-based pyramid shown here there is an axis of symmetry along the line through the vertex and the centre of the base. Around this axis the shape has rotational symmetry of order 4, since it can occupy four different positions within the same space.

EXERCISE 8

What are the symmetries of (i) a teacup, (ii) a bar of chocolate?

OUTLINE SOLUTIONS TO EXERCISES

S1

There will be 8 exterior angles each measuring $360 \div 8 = 45°$. Hence the interior angle will be $180 - 45 = 135°$.

S2

Since A and B are both points of circles where there is a tangent, there is a radius from each point to the respective centres. Since these radii arc to tangents they will both be perpendicular to the line AB, hence AD and BC are parallel.

(i)   So $ABCD$ is a trapezium.

(ii)   Since the two right angles must remain right angles, the only time the opposite angles will add up to 180 is when all four are right angles. Hence, when $ABCD$ is a rectangle; that is, when the circles are the same size.

S3

(i)   You should now consider the full size box to give: $a = 20$ cm, $b = 10$ cm, $c = 5$ cm and $d = 5$ cm. (The correct units here are important.)

(ii)   Smallest rectangle which you would need would have to be $(20 + 5 + 5)$ by $(10 + 5 + 5)$ or 30 cm by 20 cm.

S4

(a)   Since the opposite sides of a parallelogram are equal, then the triangles ABC and AFC will have all three sides the same as each other to give congruent triangles ABC and CFA, since angle BAC is equal to angle ACF.

(b)   Similar triangles will be AFC and AED.

S5

(i)   A possible answer could be:

(ii)   You will find that you can draw hexagons with 2, 3 and 6 lines of symmetry, but not 4 or 5. (In other words, the factors of 6 can be used.)

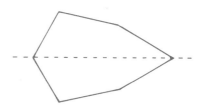

S6

(a)   Your lines of symmetry should be drawn to produce:

(You would lose marks for putting too many lines in, or missing lines.)

(b)   The only two letters are **H** and **S**.

S7

It will have just two. Both cutting the shape through the vertex and bisecting the opposite sides of the base.

S8

(i)   Teacup has one plane of symmetry only.

(ii)   Chocolate will have two planes of symmetry and one axis of symmetry.

E X A M   T Y P E   Q U E S T I O N S

Q1

This net can be folded to make a triangular prism. Which letter will point *J* join?

(SEG)

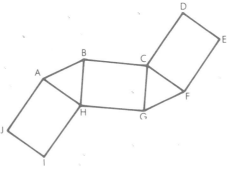

> Remember, a diagram will very often help you here.

Q2

One angle of a cyclic quadrilateral is 50°.

(a)   Write down one set of possible sizes of the other three angles.

(b)   Could these angles also be the angles of a trapezium?

Q3

In the diagram *RST*, *RV* and *TU* are tangents to the circle at *S*, *V*, and *U* respectively. Calculate the size of angle *VSU*.   (SEG)

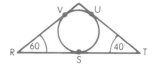

Q4

The diagram shows a triangular pyramid (tetrahedron), each of whose faces is an equilateral triangle of side 2 cm. Sketch two nets of the tetrahedron which are not congruent.   (LEAG)

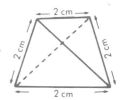

Q5

Here is a set of shapes.

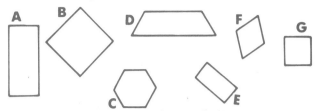

Put the letter belonging to each shape in the correct place in the diagram below.

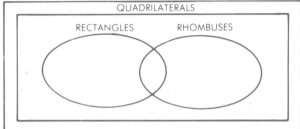

Q6

(i)   Describe the symmetries of this wallpaper pattern with reference to points A, B and C.

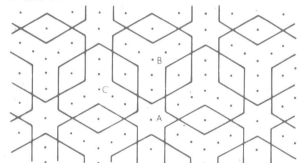

(ii)   Shade   a   rhombus already outlined and state the size of each of its angles.

(NEA)

Q7

O is the centre of the circle. Calculate the size of the angle OBC.   (LEAG)

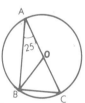

Q8

This shows a symmetrical pattern for a medallion to be made up. $ABCD$ is a kite with all sides tangential to the circle at P, Q, R and S. Calculate the size of angles (a) $\angle SMP$, (b) $\angle LRQ$.

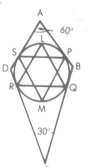

Q9

This is a design for a 'big wheel' at a fun fair. The framework is made from two regular 9-sided polygons. The corners are joined to the centre by straight struts.

Calculate the angle a and the angle b.
(Do not try to measure the drawing.)

(MEG)

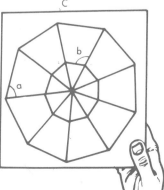

Q10

Ian has some regular hexagonal paving stones and some square paving stones, each with 30 cm sides. He needs to fit paving stones together to completely cover an area of garden.

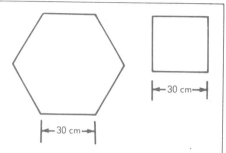

(a) What is the size of one interior angle of
   (i)   the square stone,
   (ii) the hexagonal stone?

(b) Explain why Ian cannot fit a combination of square and hexagonal stones together to cover the area of garden.

(c) Ian thinks he might make his own **regular-shaped** stone to fit with one square and one hexagonal stone.
   (i)   What would be the size of each angle of his stone?
   (ii) How many sides would his stone have?

(d) Ian decides his stone would be too large and cuts some of his hexagonal stones into equilateral triangular stones.
   (i)   Show in a sketch how he can now fit together two squares, one hexagon and one equilateral triangle at any one point.
   (ii)   Calculate the height of each equilateral triangle stone (not the thickness of the stone).
   (iii) Calculate the area covered by the arrangement in part (d)(i).

Q11

In this framework, the lengths AC and CD are equal. Angle $ABC = 71°$, angle $BAC = 59°$.

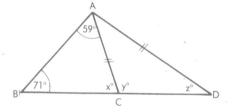

Calculate the angles marked $x°$, $y°$, $z°$.

(MEG)

Q12

The net is folded along the dotted lines to form a solid. How many edges has the solid?

(NEA)

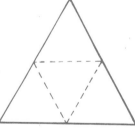

Q13

$\angle ABE = \angle ACD = 90°$.
$AC = 4.00$ m, $BC = 2.50$ m, $CD = 3.00$ m.
Calculate BE.

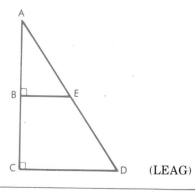

(LEAG)

Q14

The diagram shows part of a floor covered with two kinds of tiles. All the tiles are regular polygons.

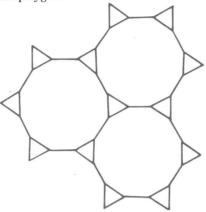

(a)   Find the sizes of the angles of the triangular tiles.
(b)   Find the sizes of the angles of the remaining tiles.   (MEG)

OUTLINE ANSWERS TO EXAM QUESTIONS

A1

F

A2

(a)   One of the angles you must give must be $180 - 50$, which is $130°$; the other two must add up to $180°$. So, possibly, the three angles could be $130°$, $110°$, $70°$. (There are lots more possibilities.)

(b)   Yes, the pairs of angles between each parallel side of a trapezium add up to $180°$, so you could have, for example, $50°$, $130°$, $50°$, $130°$, which would give a trapezium.

A3

Since $RV = RS$, then $\angle VSR = \angle SVR = (180 - 60) \div 2 = 60°$.
Similarly $\angle UST = 70°$ and hence $\angle VSU = 180 - (60 + 70) = 50°$.

A4

One answer is shown in the diagram, but there are other possible ways to do this.

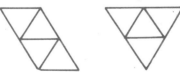

A5

Your answer should place the letters as shown on this Venn diagram.

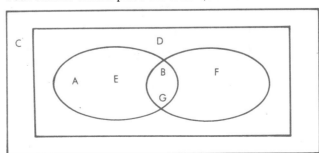

A6

(i) There are two lines of symmetry, one line being AB, the other being perpendicular to AB at point A. The shape has rotational symmetry of order 3 about the point A, or the point B or point C.

(ii) Your rhombus will have each side the length represented by 3 dots, and the angles will be 120°, 60°, 120° and 60°.

A7

Either $90 - 25$ or $(180 - 50) \div 2$ to give 65°.

A8

You must first mark in the centre of the circle, O.

(a) Since AS and AP are tangents, the right angles at S and P can be seen. This gives $\angle SOP$ to be $180 - 60$ which is 120°. Hence $\angle SMP$ is half of this, which is 60°. So $\angle SMP = 60°$.

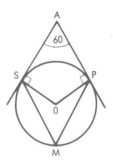

(b) For a similar reason as in part (a), we can find $\angle RLQ$ to be half of $(180 - 30)$, which will be 75°. We are told the shape is symmetrical and the only possible symmetry for this shape is a line of symmetry down the line $ALMC$, so making triangle LRQ isosceles, with $\angle LRQ$ equal to $\angle LQR$. Hence $\angle LRQ$ will be half of $(180 - 75)$ which is $52\frac{1}{2}°$.

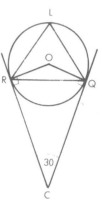

A9

The angle at the centre will be $360 \div 9 = 40°$. Hence angle a will be $\frac{1}{2}(180 - 40) = 70°$. Angle b will be $180 - 70 = 110°$.

A10

(a) (i) Size of one angle of square = 90°

(ii) Size of one angle of hexagon = $180 - (360/6) = 120°$

(b) At any point total angle = 360°

Squares cannot meet hexagons since no combination of 90 and 120 gives 360.

(c) (i) Angle = $360 - 90 - 120 = 150°$

(ii) Exterior angle = $180 - 150 = 30°$

Number of sides = $360/30 = 12$

(d)

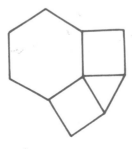 OR

A11

$x = 180 - (71 + 59) = 50$

$y = 180 - 50 = 130°; \quad z = \frac{1}{2}(180 - 130) = 25$

A12

6

A13

AB will be $4 - 2.5 = 1.5$ m

ABE and ACD are two similar triangles in the ratio 1.5 to 4.

Hence $BE = \dfrac{1.5 \times 3}{4} = 1.125$ m.

A14

(a) 60°;   (b) there are a number of ways of doing this, but the simplest way is $\frac{1}{2}(360 - 60) = 150°$

G R A D E   C H E C K L I S T

FOR A GRADE F

You should know:

the meaning of the words point, line, side, angle, acute, obtuse, right-angled, parallel, perpendicular, centre, radius and diameter;

the angle proportions of a triangle and a quadrilateral;

you should understand:

what is meant by line symmetry and rotational symmetry;

and be able to:

recognise, name and describe plane figures such as triangle (isosceles, equilateral, right-angled), square, rectangle, parallelogram, kite, quadrilateral and circle;

use a protractor to measure and draw acute and obtuse angles;

draw lines of symmetry in given shapes;

recognise, name and describe solid shapes such as cube, rectangular block, sphere, cylinder, cone and pyramid;

draw the net of a simple solid shape.

FOR A GRADE C

You should also be able to:

decide whether two shapes are congruent and give suitable explanations;

use properties of angles at a point on a straight line, and angles formed by transversal and parallel lines;

identify and use the symmetrical properties of regular and non-regular polygons;

use properties of tangent, radius and angles of a semi-circle;

sketch simple 3D shapes.

FOR A GRADE A

You should also know:

what axes of symmetry are, and what plane symmetry is;

what a cyclic quadrilateral is;

the properties of angles in a trapezium;

the properties of angles in a circle from the same chord.

STUDENT'S ANSWER - EXAMINER'S COMMENTS

QUESTION

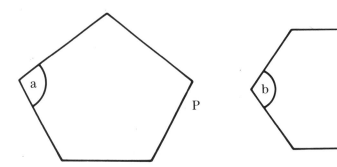

Diagrams are not
drawn to scale

The drawings show two different tiles P and H.

Tile P is in the shape of a regular 5-sided polygon.
Tile H is in the shape of a regular 6-sided polygon.

(i) Calculate the size of the angles marked 'a' and 'b'.

(ii) Explain why tiles in the shape of a 6-sided regular polygon will fit together on a floor without any gaps between them whereas tiles in the shape of a 5-sided regular polygon will not.

Answers:

(i) ANGLE a = 360° (SUM OF INTERNAL ANGLES) ÷
6 (NO. OF REGULAR ANGLES IN P) = 72°

ANGLE b = 360° (SUM OF INTERNAL ANGLES)
÷ 6 (NO. OF REGULAR ANGLES IN H) = 60°

❝A poor, thoughtless answer. The *exterior* angle has been calculated in each case instead of the *interior*.❞

(ii) THE 5-SIDED REGULAR POLYGON WILL NOT
FIT TOGETHER ON A FLOOR WITHOUT ANY
GAPS BECAUSE THE INSIDE ANGLES (108°)
WILL NOT DIVIDE INTO 360° PRECISELY
(TO 2 SIG. FIG.) WHEREAS THE SIX-
SIDED REGULAR POLYGON'S INSIDE
ANGLES (120°) WILL.

❝a good answer❞

❝A shame that this candidate would have lost a lot of marks because of not knowing what the interior and exterior angles were.❞

G E T T I N G   S T A R T E D

This chapter is all about *measuring* length, area and volume from given facts. There are a lot of formulae involved that you do not need to learn since the Examination Board will either give you a formula sheet with the formulae on it or the formulae will be stated on your examination paper. However, you do need to be *familiar* with the formulae and to be confident in *using* them. If you *do* learn a formula then it will help to make you quicker and more confident in what you are doing. The chapter brings together many ideas you will already have met in number, algebra and approximation. Although this is a small chapter many questions will be set on this topic in the examination.

USEFUL DEFINITIONS

Perimeter	length round all the outside of a flat shape
Area	flat space included in a boundary, measured in squares
Arc	part of the circumference of a circle
Volume	space inside a three-dimensional shape, measured in cubes
Hypotenuse	the longest side of a right-angled triangle
Opposite	the side of a triangle opposite to the angle concerned
Adjacent	the side of a triangle next to the angle concerned and the right angle
Elevation (angle of)	the angle measured above the horizon.
Depression (angle of)	the angle measured below the horizon.

ESSENTIAL PRINCIPLES

1 ▷ PERIMETER

The perimeter is the total outside length of any flat shape. The perimeter of a circle, known as the *circumference*, can be found by the formula:

circumference = π multiplied by the diameter,

where π can be taken to be 3, or 3.1 or 3.14, or found more accurately by pressing the π button on a calculator.

WORKED EXAMPLE 1

Which has the longer perimeter; a rectangle measuring 6 cm by 2 cm or a circle of diameter 5 cm?

The perimeter of the rectangle will be 6 + 2 + 6 + 2 = 16 cm.
The perimeter (circumference) of the circle is approximately 3.14 × 5 = 15.7 cm, so the perimeter of the rectangle is the longer.

EXERCISE 1

The diameter of the earth is approximately 7900 miles.

(a) Freda went half way round the world to visit her aunt. How far had she travelled?
(b) Mr Graves travelled round the world in 80 days. How many miles would he average a day?

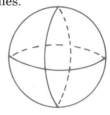

2 ▷ AREA

Area is the amount of space inside a flat 2D shape, and is measured in squares, e.g. square centimetres or square yards.

The area of a rectangle can be found with the formula

area = length × breadth.

The area of a triangle can be found with the formula

area = $\frac{1}{2}$ of base length × height.

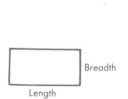

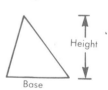

WORKED EXAMPLE 2

Find the total surface area of the cuboid illustrated.

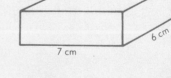

Each face is a rectangle, and the opposite faces are equal. So there will be a total surface area of

2 × [(7 × 6) + (7 × 3) + (6 × 3)]
  = 2 × (42 + 21 + 18),

which is 2 × 81, or 162 square centimetres, written as 162 cm².

UNUSUAL SHAPES

For awkward shapes we can calculate the area by placing a suitable squared grid over the shape and counting the whole squares and estimating what the bits add up to.

WORKED EXAMPLE 3

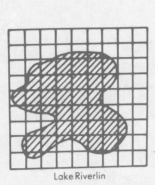
Lake Riverlin

Here is a map of Lake Riverlin, which has a scale of 1 square unit to 1 square kilometre. What is the surface area of Lake Riverlin?

Count the whole squares – you should get 13. Now go through the smaller bits and estimate how many whole squares they would add up to. This should be close to 14, giving a total of 27 square units. Hence Lake Riverlin will have a surface area of 27 square kilometres.

PARALLELOGRAM

The area of a *parallelogram* is found by the formula

area = base length × height.

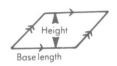

TRAPEZIUM

The area of a *trapezium* is found by multiplying the height by the average length of the parallel sides. This is often written as:

$$\text{area} = \frac{h}{2}(a + b) \quad \text{or} \quad h\left(\frac{a + b}{2}\right)$$

both of which are the same formula.

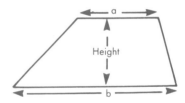

WORKED EXAMPLE 4

Which of the two shapes below has the larger area?

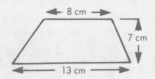

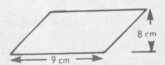

The left-hand shape is a trapezium with an area of $\frac{7}{2}(8 + 13) = 73.5$ cm^2. The right-hand shape is a parallelogram with an area of $8 \times 9 = 72$ cm^2. Hence the trapezium has the larger area.

CIRCLE

The area of a *circle* is found by the formula

area = π × (radius)^2

where you should use the π on your calculator.

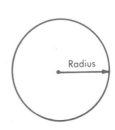

SECTOR

The area of a *sector* of angle x is found by finding the fraction $\dfrac{x}{360}$ of the whole circle area, i.e.

$$\text{sector area} = \frac{x}{360} \times \pi \times (\text{radius})^2.$$

WORKED EXAMPLE 5

VACMAN, as shown below, starts as a sector of a circle with a radius of 3 cm; each smaller circle is 1 cm in radius. Each time VACMAN eats a circle he increases in area by the same area as that of the circle he eats.

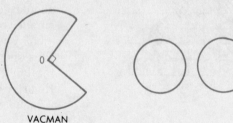

VACMAN

(a) Calculate the area of VACMAN before he starts eating.
(b) Calculate the area of one small circle.
(c) After VACMAN has eaten 14 small circles, calculate (i) his new area and (ii) his new radius assuming he stays a similar shape.

> A typical easy error here is to think of the sector angle as 90°, so be careful.

(a) This is a sector of angle 270° and radius 3 cm, so its area is given by

$$\frac{270}{360} \times \pi r^2 = \frac{270}{360} \times \pi \times 9$$

which rounds to 21.2 cm², giving the area of VACMAN.

(b) One small circle area will be πr^2 where $r = 1$, hence each is 3.1 cm² (or π cm²).

(c) (i) VACMAN's new area will be his old area + 14π, so being as accurate as possible you should use $(21.20575 + (14 \times \pi))$, which is 65.2 cm² (rounded off).

   (ii) If the new area is 65.2, then where r is the radius we shall have the equation

$$65.2 = \frac{270}{360} \times \pi r^2, \quad \text{rearranged to give} \quad r^2 = \frac{65.2 \times 360}{270 \times \pi},$$

so using the accurate area of 65.2 (before it was rounded off), this will give $r^2 = 27.66\ldots$ (do not round off yet), which will give $r = 5.3$ cm (rounded off) as VACMAN's new radius.

EXERCISE 2

A triangle, parallelogram and circle all have the same height of 6 cm, and the same area. What are the base lengths of (i) the triangle and (ii) the parallelogram. (Give answers to one decimal place.)

It also follows that a *length of arc*, as seen in the diagram, is found by the fraction $\dfrac{x}{360}$ of the whole circumference. Hence, the arc length of a sector of angle x is given by

$$\text{arc length} = \frac{x}{360} \times \pi \times \text{diameter}.$$

Since, the diameter is $2 \times$ radius, it follows that

$$\text{arc length} = \frac{x}{360} \times 2\pi r.$$

<table>
<tr><td>**WORKED**
EXAMPLE 6</td><td>

Find the arc length AB in the sector
AOB.

Arc length is given by
$$\frac{70}{360} \times 2\pi \times 6 = 7.3 \text{ cm (rounded off)}.$$

</td><td>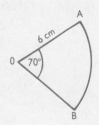</td></tr>
</table>

EXERCISE 3

A pendulum swings through an angle of $40°$. How much further does the end of a 30 cm pendulum move, than a point half way down it?

4 ⟩   **VOLUME**   *Volume* is the amount of space inside a 3D shape, and is measured in cubes, e.g. cubic millimetres and cubic metres. The volume of a cuboid is found by multiplying length by breadth by height. For example, the volume of the cuboid in the diagram is $7 \times 6 \times 3$, which is 126 cubic centimetres, written as 126 cm^3.

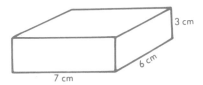

WORKED
EXAMPLE 7

(a) How many boxes are in this pile?

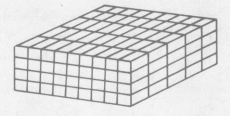

(b) Fred put some of the boxes from the pile into this trolley. How many of the boxes altogether will fit into this trolley?

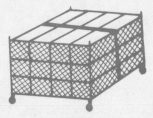

(c) Fred was told to move all the boxes in the pile, using the trolley. How many times must he fill the trolley in order to move them all?

(a) The number of boxes will be found by finding how many boxes there are along the length, breadth and height of this pile and multiplying together to give $9 \times 5 \times 4$, which gives a total of 180 boxes in the pile.

(b) Using the same method as before to see how many boxes will fit into length, breadth and height, you should get $4 \times 2 \times 3$, which is 24 boxes which Fred can fit into his trolley.

(c) The number of journeys Fred had to make to move all the boxes in the pile will be found by dividing 180 by 24, which gives 7.5. So your answer should be 8 journeys, or 7 full loads and a small one. 7.5 on its own would not get full marks.

PRISMS

The volume of any *prism* is found by multiplying the area of the regular cross section (which is the same as the end!) by the length of the shape.

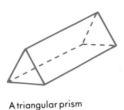

A hexagonal prism

A triangular prism

So, for example, in the prisms shown here, the first thing to calculate is the area of the end, then multiply by the length of the shape.

WORKED EXAMPLE 8

Calculate the volume of the prism. Volume is area of triangular end × length which is

$\frac{1}{2} \times 3 \times 4 \times 5 = 30 \text{ cm}^3$

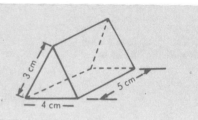

3 cm
5 cm
4 cm

5 ▷ MEASURING SOLIDS

You should be familiar with the following solid shapes and their mensuration.

CYLINDER

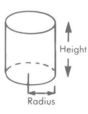

Height

Radius

The *volume* of a *cylinder* is found by multiplying its base area by its height, so giving

$$\text{volume} = \pi r^2 h$$
$(r = \text{radius}; h = \text{height})$

The *curved surface area*, that is just the curved part of the cylinder, is found by the formula

$$\text{curved surface area} = \pi D h$$
$(D = \text{diameter}; h = \text{height})$

The *total surface area* of a cylinder then will be found by

$$\text{total surface area} = \pi D h + 2\pi r^2$$

SPHERE

The *surface area* of a *sphere* of radius r is given by the formula:

$$\text{surface area} = 4\pi r^2$$

The *volume* is given by:

$$\text{volume} = \frac{4}{3}\pi r^3$$

PYRAMID

The *volume* of a *pyramid* is found by multiplying its base area by one-third of its height

$$volume = \frac{h}{3} \times (\text{base area}),$$

where $h = \text{height}$.

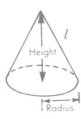

CONE

Where a *cone* of height h has a base radius r and a slant height of l, then:

the *curved surface area* is given by:

$$\text{CSA} = \pi r l$$

the *volume* is given by:

$$\text{volume} = \tfrac{1}{3}\pi r^2 h$$

EXERCISE 4

A gold sphere, of radius 10 cm, was melted down and made into a cone whose height is the same as the sphere's diameter. What is the radius of the base of the cone?

6 ▷ SOLUTION OF TRIANGLES

PYTHAGORAS

You ought to know the rule of Pythagoras, which says that *in a right-angled triangle the squares of the two smaller sides add up to the same as the square of the longest side* (hypotenuse), i.e.

$$a^2 + b^2 = c^2$$

You will need to use this rule in two ways, illustrated in the following example.

WORKED EXAMPLE 9

Kate has a ladder that will reach 12 feet up the wall when the bottom is 5 feet away from the wall.
(a) How long is the ladder?
(b) How far up the wall will this ladder reach when its bottom is 4 feet away from the wall?

Since the diagram illustrates that we have a right-angled triangle, we can use the rule of Pythagoras.
(a) In the triangle made with the ladder, the wall and the floor, the two small sides are 5 and 12. Hence, if the hypotenuse is called x then $x^2 = 5^2 + 12^2 = 25 + 144 = 169$. Hence $x = \sqrt{169} = 13$, so the ladder is 13 feet long.
(b) In this part we know the hypotenuse and need to find a small side. Using the same rule and calling the unknown small side y,

we have $\qquad 4^2 + y^2 = 13^2$

that is $\qquad 16 + y^2 = 169$

giving $\qquad y^2 = 169 - 16 = 153$

hence $\qquad y^2 = \sqrt{153} = 12.4$ (rounded off).

So the ladder will reach 12.4 feet up the wall.

Be carefull to round off properly, not doing so throws marks away.

EXERCISE 5

In a right-angle triangle two sides are known to be 4 cm and 7 cm. What two possible areas could the triangle have?

7 ▷ TRIGONOMETRY

You have probably spent quite a lot of time on trigonometry already, but here are the main facts that you ought to be familiar with.

In any right-angled triangle we call the long side, which is always opposite the right angle, the *hypotenuse*. Then, depending on which angle of the triangle we are finding or going to use, we name the other two sides. The side opposite the angle we call *opposite* and the one next to both the angle under consideration and the right angle, we call the *adjacent*, as illustrated.

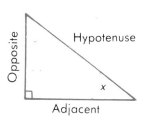

Then, for the given angle x

$$\text{tangent} = \frac{\text{opposite}}{\text{adjacent}} \qquad \text{sine} = \frac{\text{opposite}}{\text{hypotenuse}} \qquad \text{cosine} = \frac{\text{adjacent}}{\text{hypotenuse}}$$

You may well be given this information on a formula sheet in the examination but not every Examination Board does this, so check it out for yourself. But in any case it's useful to remember this information, and one way of doing it is to learn a sentence to help. For example, we can abbreviate the formulae to T = O/A, S = O/H, C = A/H, which can be put into a sentence such as 'Tommy On A Ship Of His Caught All Herring'. Of course, you can make up one of your own that *you* find easier to remember.

This information is used in two ways. Firstly, to find the size of angles, and, secondly, to calculate lengths of triangles.

TO FIND ANGLES

If you are finding an *angle* in a right-angled triangle and you know all three sides, then you have the choice of three ways to find the size of the angle. However, usually you will only know two sides, and therefore only one way is suitable. Look at the following examples, where in each right-angled triangle we are calculating the size of the angle x.

WORKED EXAMPLE 10

We look first to see which sides we know. These are the 'opposite' and the 'adjacent', hence we need to calculate 'tangent'. Using the previous information:

$$\text{tangent}\, x = \frac{\text{opposite}}{\text{adjacent}} = \frac{7}{5} = 1.4$$

We now need the angle that has a 'tan' of 1.4. It is best to do this on the calculator by obtaining 1.4 in the display, finding and pressing the $\tan^{-1}$ button (often by pressing 'INV' first then 'tan'), and hence obtaining an answer that will round off to 54.5°.

WORKED EXAMPLE 11

This time we have 'opposite' and 'hypotenuse' which leads us to sine. Hence we can say:

$$\sin x = \frac{\text{opposite}}{\text{hypotenuse}} = \frac{3}{8} = 0.375$$

and again find $\sin^{-1}$ to press on the calculator, giving 22.0°.

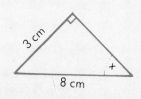

<table>
<tr><td>

WORKED EXAMPLE 12

</td><td>

Here we are given the 'adjacent' and the 'hypotenuse' which leads us to cosine. We can say:

$$\text{cosine } x = \frac{\text{adjacent}}{\text{hypotenuse}}$$

$$= \frac{9}{11} = 0.81818\ldots$$

(leave it all in the calculator)

Press $\cos^{-1}$ to obtain $35.1°$.

</td><td>

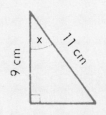

</td></tr>
</table>

NOTE

If you do not possess a calculator with trigonometrical functions, then you will have to use trigonometrical tables to find these figures, which will be a serious disadvantage to you.

TO FIND LENGTHS

In a right-angled triangle, once you are told one of the other angles, say $25°$, then you are in a position to find the other angle. Here it will be $90 - 25$ which is $65°$. So really again you often have a choice of methods to use to find a missing *length*, but as far as possible always try to use the information given to you in the first place (it is more likely to be correct!). Look at these following examples of finding missing lengths.

<table>
<tr><td>

WORKED EXAMPLE 13

</td><td>

The given side is 'adjacent', and the side we are finding, y, is the 'opposite', so we shall use tangent $25°$ to give:

$$\text{tangent } 25 = \frac{\text{opposite}}{\text{adjacent}} = \frac{y}{5}$$

So rearrange to give

$$y = 5 \times \text{tangent } 25.$$

Put 25 into the calculator and press 'tan' to find the tangent of 25. Now multiply by 5 and round off to get 2.3 cm.

</td><td>

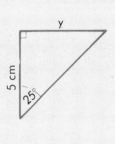

</td></tr>
</table>

<table>
<tr><td>

WORKED EXAMPLE 14

</td><td>

The two sides we are involved with here are 'opposite' and 'hypotenuse', hence we need to use sine $36°$. This will give us:

$$\text{sine } 36 = \frac{\text{opposite}}{\text{hypotenuse}} = \frac{y}{8}$$

So rearrange to give

$$y = 8 \times \sin 36.$$

Put 36 into the calculator and press 'sin'. Now multiply by 8 and round off to 4.7 cm.

</td><td>

</td></tr>
</table>

<table>
<tr><td>

WORKED EXAMPLE 15

</td><td>

We are involved here with cosine, and can write:

$$\text{cosine } 65 = \frac{y}{15}$$

Rearrange to give

$$y = 15 \times \text{cosine } 65,$$

which gives us $y = 6.3$ cm.

</td><td>

</td></tr>
</table>

NOTE

If it's the *hypotenuse* we are being asked to find, then we have to be careful. Look at the following example.

WORKED EXAMPLE 16

We can recognise the sine situation and write down:

$$\text{sine } 57 = \frac{\text{opposite}}{\text{hypotenuse}} = \frac{8}{y}$$

which will rearrange to give

$$y = \frac{8}{\sin 57}$$

This is where you need to be specially careful, for different calculators can do this in many different ways. One way that will work on all scientific calculators is to put 57 into the display, press 'sin' and put the result into memory, then put 8 into the display and divide this by the 'memory recall' to give a rounded answer of 9.5 cm. But do look to see how *your* calculator can do this as simply as possible.

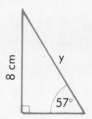

3D SITUATIONS

At the higher levels of GCSE you will be expected to be able to use the trigonometry you have learned to solve three-dimensional problems. This will often involve what we call 'dropping perpendiculars', that is to say, if something was to fall to the ground from any point above the ground, it would fall in a vertical line, perpendicular to the horizontal.

In any 3D situation that you are required to work with, it is vital that you are able to 'see' the right angles and use them.

WORKED EXAMPLE 17

The diagram illustrates a prisoner's escape hole just on the edge of the perimeter fence. There was one lookout on a tower 8 yards high due north, and another lookout on a tower 10 yards high due west. Both lookouts were 18 yards away. The prisoner tried to escape one night when visibility was 20 yards.

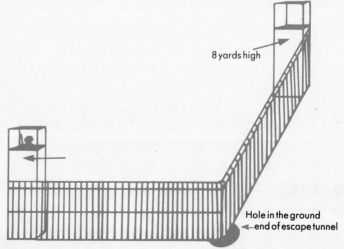

(a)   Which of the two lookouts could see the prisoner escape?

(b)   What visibility would you need for one lookout just to be able to see the other?

(a) Looking at the situation of the northern lookout we have the triangle shown above. The distance from the lookout to the prisoner's hole is the hypotenuse, which is found by the rule of Pythagoras:

$$x^2 = 18^2 + 8^2 = 388,$$

hence $x = 19.7$ yards. This distance is less than 20 yards, so the guard in this lookout tower can see the escape hole. A similar situation from the viewpoint of the other lookout gives a right-angled triangle with a solution of

$$x^2 = 18^2 + 10^2 = 424,$$

hence $x = 20.6$ yards. This distance is greater than 20 yards, so the guard in this lookout tower cannot see the escape hole.

(b) Drawing a line from one lookout to the other gives us the diagram below.

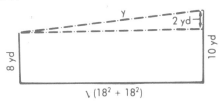

The distance between the lookouts is found by considering the right-angled triangle formed between the foot of each lookout tower and the escape hole to give $\sqrt{(18^2 + 18^2)}$. So the actual distance between the two lookouts will be found by the rule of Pythagoras:

$$y^2 = \left(\sqrt{(18^2 + 18^2)}\right)^2 + 2^2$$

which is 652. Hence y is $\sqrt{652}$ which is 25.53. Therefore a visibility of 26 yards will just allow one lookout to see the other.

EXERCISE 6

This diagram represents the roof of a barn with a rectangular base and 'isosceles triangular' ends both sloping in at the same angle.

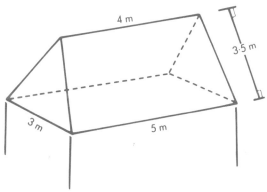

Calculate the slope made with the horizontal of (i) the rectangular faces of the roof; (ii) the triangular faces of the roof.

SINE RULE

In a few syllabuses trigonometry is extended to solve non-right-angled triangles with the use of the *sine rule* and the *cosine rule*. But do check to see if this is necessary for the syllabus you are taking, since it is not in many. However, it could be interesting to look at the topic even if you do not need to do it.

Draw a triangle (obtuse if you wish), and label it as in the diagram. Notice how the small letters come opposite the capital letters. Then measure each angle and the length of each side. You can now check for yourself the *sine rule*, which will work for **any** triangle.

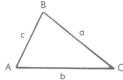

$$\frac{a}{\sin A} = \frac{b}{\sin B} = \frac{c}{\sin C} \quad \text{OR} \quad \frac{\sin A}{a} = \frac{\sin B}{b} = \frac{\sin C}{c}$$

WORKED EXAMPLE 17

Find the marked length x.

We can apply the sine rule to give:

$$\frac{x}{\sin 110} = \frac{8}{\sin 40},$$

which gives

$$x = \frac{8 \sin 110}{\sin 40} = 11.7 \text{ cm.}$$

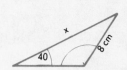

WORKED EXAMPLE 18

Find the marked angle θ.

We apply the sine rule to give

$$\frac{\sin \theta}{3} = \frac{\sin 130}{5},$$

which gives

$$\sin \theta = \frac{3 \sin 130}{5} = 0.4596(\ldots)$$

giving $\theta = 27.4°$.

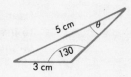

COSINE RULE (1)

This is used when we know all three sides of a triangle and we wish to find an angle.

If we wish to find angle A then:

$$\cos A = \frac{b^2 + c^2 - a^2}{2bc}$$

(note which side comes last).

For angle B:

$$\cos B = \frac{a^2 + c^2 - b^2}{2ac}$$

Try to write the rule for angle C yourself.

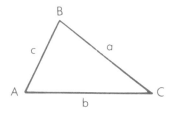

Find angle θ.

We apply the cosine rule to give

$$\cos \theta = \frac{5^2 + 7^2 - 6^2}{2 \times 5 \times 7} = 0.5428(\ldots)$$

giving $\theta = 57.1°$.

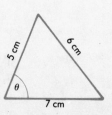

COSINE RULE (2)

If we are given information about two sides and the included angle, then we can find another side by turning the cosine rule round to give

$$a = \sqrt{(b^2 + c^2 - 2bc \cos A)} \quad \text{OR} \quad b = \sqrt{(a^2 + c^2 - 2ac \cos B)}$$

Try to write the rule for c yourself.

WORKED EXAMPLE 20

Find the marked side x.

We apply the cosine rule to give

$$x = \sqrt{(5^2 + 8^2 - 2 \times 5 \times 8 \cos 47)}$$

$$= 5.9 \text{ cm.}$$

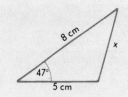

NOTE

Remember when to use each one by a simple rule: 'When two sides involved make the angle cosy, use the cosine rule – otherwise use the sine rule.'

AREA

A simple way to find the *area of a triangle*, if you know two sides and the included angle, is to use:

$$\text{Area} = \tfrac{1}{2} ab \sin C$$

OR $\text{Area} = \tfrac{1}{2} bc \sin A$

Try to write the rule if it is angle B which you are given.

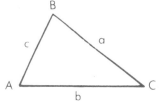

WORKED EXAMPLE 21

Find the area of the triangle.

Use the area sine formula:

$$\text{Area} = \tfrac{1}{2} \times 7 \times 9 \times \sin 63$$

$$= 28 \text{ cm}^2.$$

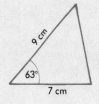

ELEVATION AND DEPRESSION

You may come across an angle defined as an 'angle of depression', which is like angle A, an angle made with a line below the horizon. Or you may come across an 'angle of elevation' which is like angle B, an angle made with a line above the horizontal.

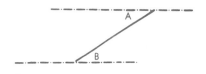

SOLUTIONS TO EXERCISES

S1

(a) The circumference of the world will be π multiplied by the diameter of the world, which will be $\pi \times 7900$. Do this on your calculator and round off to 24800 miles. Hence halfway round the world will be approximately half of 24800 miles, which is 12400 miles.

(b) Divide the most accurate figure you had for the circumference around the world by 80, then round off to 310, giving the average day's journey by Mr Graves to be 310 miles.

S2

Area of circle $= \pi r^2 = \pi \times 3^2 = \pi \times 9 = 28.27(\ldots)$. Put this accurate calculator display into the memory, and we'll use it rather than the 28.27.

(i) From area of triangle $= \frac{1}{2}$ base length $\times$ height, we get the equation $28.27(\ldots) = \frac{1}{2} \times$ length $\times 6$, which re-arranges to give

$$\text{length} = \frac{2 \times 28.27(\ldots)}{6} = 9.4 \text{ cm}$$

(rounded off).

(ii) From area of parallelogram = base length $\times$ height we get the equation $28.27(\ldots) = \text{length} \times 6$, which solves to give

$$\text{length} = \frac{28.27(\ldots)}{6} = 4.7$$

(rounded off).

S3

Arc length of the end of the pendulum is given by

$$ED = \frac{40}{360} \times 2\pi \times 30 = 20.94.$$

Length of the arc in the middle is given by

$$MN = \frac{40}{360} \times 2\pi \times 15 = 10.47.$$

The answer is that the pendulum moves 10.5 cm more (rounded off).

S4

Calculate the volume of the sphere, given by $\frac{4}{3}\pi r^3$ where $r = 10$, and this will be 4189 (rounded, but keep the accurate value in the memory of your calculator to use later). The volume of a cone is given by $\frac{1}{3}\pi r^2 h$, so when the volume is as above and the height equal to 20 cm, we have $4189 = \frac{1}{3}\pi r^2 \times 20$ which can be rearranged to give

$$r^2 = \frac{4189 \times 3}{\pi \times 20} = 200$$

(using the accurate value of 4189). Hence $r = \sqrt{200}$, so the radius of the base of the gold cone will be 14.1 cm.

S5

The two possible triangles are as follows:

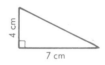

The left-hand triangle has an area of $\frac{1}{2} \times 4 \times 7 = 14$ cm². We need to find the smaller missing side in the right-hand triangle. This length is given by $\sqrt{(7^2 - 4^2)} = \sqrt{33} = 5.7(\ldots)$. The area of this triangle is

$$\frac{1}{2} \times 4 \times 5.7 \, (\ldots) = 11.5 \text{ cm}^2$$

(rounded off). So, the possible areas are 14 cm² and 11.5 cm².

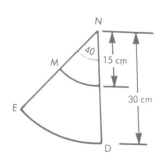

S6

(i)   From any point along the 4 m edge of roof top imagine a perpendicular dropped to the base of the roof and a line drawn perpendicular to the 5 m edge. This gives a right-angled triangle as shown, with x the angle we are asked for.

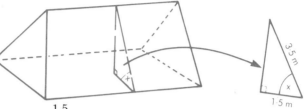

This is found with $\cos x = \dfrac{1.5}{3.5}$, hence x is 64.6°, rounded to one decimal place.

(ii)   From the triangle in (i), you can calculate the height of the roof by the rule of Pythagoras.

Height$^2 = 3.5^2 - 1.5^2 = 10$,   hence height is $\sqrt{10}$ (no need to calculate yet). Now, from the very edge of the 4 m roof line, drop a perpendicular to the roof base, and imagine a line drawn perpendicular to the 3 m edge to give the right-angled triangle seen here.

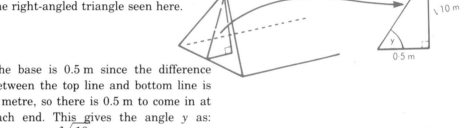

The base is 0.5 m since the difference between the top line and bottom line is 1 metre, so there is 0.5 m to come in at each end. This gives the angle y as: $\tan y = \dfrac{\sqrt{10}}{0.5}$, so y will be 81.0°.

EXAM TYPE QUESTIONS

Q1

Here are some solids made from centimetre cubes.
Find the total surface area of each solid.   (NEA)

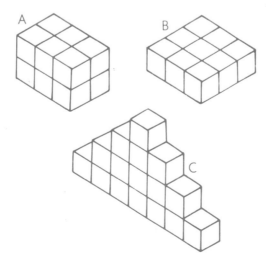

Q2

The instructions for erecting a greenhouse say: 'First make a rectangular base 2 m by 3.5 m. To check that it is rectangular measure the diagonal. It should be about 4 m.'

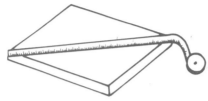

Using Pythagoras' theorem, explain why the diagonal should be about 4 m.   (SEG)

Q3

In the parallelogram $ABCD$, BE is perpendicular to AD, angle $A = 70°$, $AB = 8$ cm and $BC = 10$ cm.

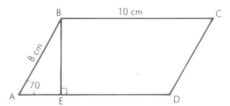

Calculate, giving your answers correct to two significant figures,
(a) the length of BE;
(b) the area of the parallelogram $ABCD$.   (MEG)

Q4

A square has area 2500 cm².
What is the length of each side of the square?

2500 cm²

(NEA)

Q5

John's cycle has wheels of radius 1 ft.

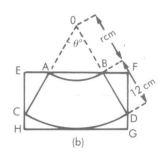

(a) Calculate the circumference of John's front wheel.
    (Either take π as 3.14 or use the π button on your calculator.)
(b) (i)   Calculate how far John has cycled when the front wheel has rotated 70 times.
    (ii)   Give this distance to the nearest hundred feet.   (SEG)

Q6

A mathematical D-I-Y enthusiast plans to re-cover a lampshade of circular cross-section. The dimensions of the lampshade are shown in figure (a). The fabric is to be cut from a rectangular piece of material, $EFGH$, to the pattern shown in figure (b). The arcs AB and CD are from separate circles with the same centre O and each arc subtends the same angle $\theta°$ at O.

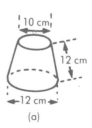

(a) (b)

(a)   Using the measurements given in figure (a) write down, as a multiple of π, the lengths of (i) arc AB, (ii) arc CD.
(b)   Using the measurements given in figure (b) write down, in terms of r, θ and π, the lengths of (i) arc AB, (ii) arc CD.
(c)   Use the results of (a) and (b) to find (i) r, (ii) θ.   (LEAG)

Q7

Two lookout posts, A and B on a straight coastline running east-west sight a ship (S) on a bearing 067° from A and 337° from B.

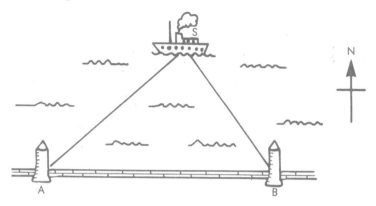

(a) Explain why angle ASB is 90°.

The distance from A to B is 5 kilometres.

(b) Calculate the distance of the ship from A.

(c) Calculate the distance of the ship from B.

The ship sails on a course such that angle ASB is always 90°.

(d) Describe the path the ship must take.

(e) What is the bearing of the ship from A (to the nearest degree) when it is 3 kilometres from it?   (SEG)

Q8

The diagram represents two fields ABD and BCD in a horizontal plane $ABCD$.

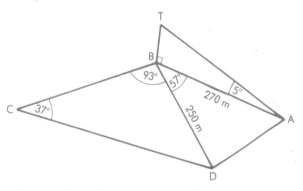

$AB = 270$ m, $BD = 250$ m,   angle $ABD = 57°$,   angle $BCD = 37°$ and   angle $CBD = 93°$.

A vertical radio mast BT stands at the corner B of the fields and the angle of elevation of T from A is 5°. Calculate, correct to three significant figures,

(a) the height of the radio mast,

(b) the length of AD,

(c) the length of BC,

(d) the area, in hectares, of the triangular field ABD (1 hectare = 10^4 square metres).   (MEG)

Q9

A, B and C are three points on the map. They all lie on the 10 metre contour line but are separated by the river estuary as shown. The map surveyor measured the straight line distance from A to B as 120 m and the angles BAC and ABC were found to be 54° and 90° respectively.

Calculate the straight line distances BC and AC.   (NEA)

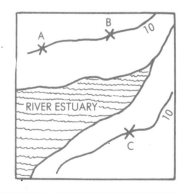

Q10

(a) Write down an expression for BD in terms of h.

(b) Write down an expression for DC in terms of h.

(c) Hence calculate h.   (LEAG)

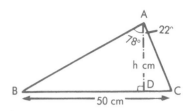

Q11

AB is a chord of a circle and AC is a diameter. The length of AB is 14 cm and the radius of the circle is 25 cm. Calculate the length of the chord BC.   (MEG)

Q12

The tables in a Burger Bar are circular, with a minor segment removed to form a straight edge. They have a diameter of 1 metre, and angle BOD is 90°. The tops are covered with formica and the perimeter is bound with thin steel strip. Calculate the area of the table top and the length of strip required.   (SEG)

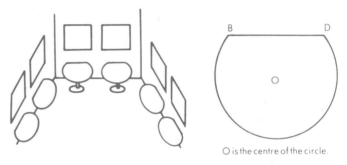

O is the centre of the circle.

Q13

A solid silver sphere has a radius of 0.7 cm.

(a) Calculate

　(i)　the surface area of the sphere,

　(ii)　the volume of the sphere.

(b) A silversmith is asked to make a solid pyramid with a vertical height of 25 cm and a square base. To make the pyramid, the silversmith has to melt down 1000 of the silver spheres. Assuming that none of the silver is wasted, calculate the total surface area of the pyramid.   (MEG)

Q14

These instructions are issued by the British Decorating Association.

Working at Heights

If a ladder is used as a means of access, it should rise to a height of 3 ft 6 in (1.074 m) above a landing place.

Ladders should be set at 75° to the horizontal.

A builder has a contract to resurface a flat roof which is 7.5 m above ground level. What is the minimum length of ladder that should be issued to the workers? (NEA)

Q15

A cone has a base radius of 5 cm and a vertical height of 12 cm. A circle is drawn on the surface at a vertical height of 6 cm. The surface below the circle is painted red. What area, in cm², is painted red? Take $\pi = 3.14$.   (MEG)

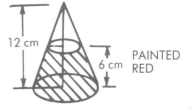

Q16

(i) In the diagram $a = 7$ and $b = 24$.
Calculate the value of c.

a	b	c
3	4	5
5	12	13
7	24	
9	40	41
11	60	61
13	84	85
15		113
17		

(ii) Use your answer to part (i) to complete the third row in the table.

(iii) By considering the patterns in the rows and the columns of the table, complete the remaining rows.

(NEA)

Q17

The shape shows the plan of a flower-bed. It is formed from four quarter circles, each radius 5 m. Calculate the perimeter of the flower-bed, taking $\pi = 3.1$.

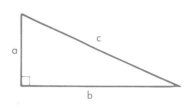

Q18

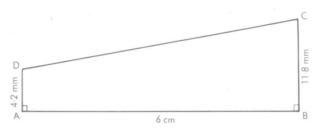

The diagram shows the cross-section $ABCD$ of a plastic door wedge.

(a) Write down, in cm, the length of BC.

(b) Calculate, in cm², the area of the cross-section $ABCD$.

(c) Given that the wedge is of width 3 cm, calculate the volume, in cm³, of plastic required to make

(i) 1 wedge,

(ii) 1 000 000 wedges.

OUTLINE ANSWERS TO EXAM QUESTIONS

A1

$A = (4 \times 6) + (2 \times 4) = 32$ cm².
$B = (2 \times 9) + (4 \times 3) = 30$ cm².
$C = (2 \times 16) + 7 + 15 = 54$ cm².

A2

Because $\sqrt{(2^2 + 3.5^2)} = \sqrt{(16.25)} = 4.03$, to two significant figure accuracy, the answer is 4.0, or 4 m.

A3

(a) $BE = 8 \sin 70 = 7.5$ cm,

(b) Area $= 10 \times 7.5 = 75$ cm².

A4

$\sqrt{2500} = 50$ cm.

A5

(a) Circumference $= \pi D = \pi \times 2 = 6.28$ ft.

(b) (i) $70 \times \pi \times 2 = 440$ ft (rounded off);   (ii) 400 ft.

A6

(a) (i)   arc $AB = \pi D = 10\pi$;   (ii)   arc $CD = \pi D = 12\pi$.

(b) (i)   arc $AB = \dfrac{\theta}{360} \times 2 \times \pi \times r$;   (ii)   arc $CD = \dfrac{\theta}{360} \times 2 \times \pi \times (r + 12)$.

(c) (i)   From $AB = 10\pi = \dfrac{\theta}{360} \times 2\pi r$;   and $CD = 12\pi = \dfrac{\theta}{360} \times 2\pi (r + 12)$

$$\frac{10\pi}{2\pi r} = \frac{\theta}{360} = \frac{5}{r} \qquad \frac{12\pi}{2\pi(r + 12)} = \frac{\theta}{360} = \frac{6}{r + 12}$$

hence $\dfrac{\theta}{360} = \dfrac{5}{r} = \dfrac{6}{r + 12}$,

hence $5r + 60 = 6r$;   giving $r = 60$ cm.

(ii)   Substitute back into one of the equations to get $\theta = 30°$.

A7

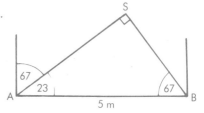

(a) There are many ways to explain this. One of them is:

    A is on a bearing of $(180 + 067) = 247°$ from S.

    B is on a bearing of $(180 + 337) - 360 = 157°$ from S.

    Hence the difference between A and B is $247 - 157 = 90°$.

(b) From $\triangle ABS$,

$AS = 5 \cos 23 = 4.6$ m.

(c) $SB = 5 \cos 67 = 2.0$ m.

(d) The ship will sail in a circle of diameter AB.

(e) $\angle SAB = \cos^{-1} \frac{3}{5} = 53°$ (to nearest degree) so the bearing will be $90 - 53° = 37°$.

A8

(a) $TB = 270 \tan 5 = 23.6$ m.

(b) Use cosine rule to give

$$AD = \sqrt{(270^2 + 250^2 - 2 \times 270 \times 250 \times \cos 57)} = 249 \text{ m.}$$

(c) Use sin rule to give $BC = \dfrac{250 \times \sin(180 - 37 - 93)}{\sin 37} = 318$ m.

(d) Use sine rule to give $A = \dfrac{1}{2} \times 250 \times 270 \times \sin 57 = 28\,300$ m².

A9

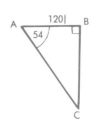

$BC = 120 \tan 54° = 206$ m.

$AC = \dfrac{120}{\cos 54} = 204$ m.

A10

(a) $h \tan 78$ or $h \times 4.7(...)$.   (b) $h \tan 22$ or $h \times 0.404(...)$.

(c) $h \times 4.7(...) + h \times 0.404(...) = 50 = h \times 5.108(...)$,

    $h = 50 \div 5.108(...) = 9.787(...)$

    round off to 9.8 cm.

A11

You should have sketched a semi-circle with a right-angled triangle ABC. Hence
$BC = \sqrt{(25^2 - 14^2)} = 20.7$ cm.

A12

Area of sector $= \dfrac{270}{360} \times \pi \,(0.5)^2$

$\qquad\qquad = 0.589$ m^2.

Area of triangle $= \frac{1}{2}(0.5) \times 0.5$

$\qquad\qquad = 0.125$ m^2.

Total area of table $= 0.714$ m^2 or 7140 cm^2.

Arc length $BD = \dfrac{270}{360} \times \pi \times 1$

$\qquad\qquad = 2.356$ m.

Straight length
$BD = \sqrt{(0.5^2 + 0.5^2)} = 0.707$ m.

Total strip needed $= 3.06$ m or 306 cm.

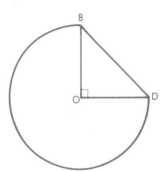

A13

(a) (i) $4\pi r^2 = 4 \times \pi \times (0.7)^2 = 6.16$ cm^2.

(ii) $\frac{4}{3}\pi r^3 = \frac{4}{3} \times \pi \times (0.7)^3 = 1.44$ cm^3.

(b) Volume of pyramid $= 1436.755$ (using the accurate answer in calculator for (a) (ii).

Base of pyramid $= \dfrac{1436.755 \times 3}{25}$ (from volume $= \frac{1}{3}$ base $\times$ height)

$\qquad\qquad = 172.4106$.

Length of base $= \sqrt{172.(\ldots)} = 13.1$ cm.

Area of one triangle face of pyramid $= \frac{1}{2} \times 13.1 \times l$ ($l =$ slant height).

Slant height $= \sqrt{\left(25^2 + \left(\dfrac{13.1}{2}\right)^2\right)} = 25.8(\ldots)$.

Area of one triangle face $= 169.3$ cm^2.

Total surface area $= (4 \times 169.3) + 172.4106 = 850$ cm^2.

A14

Length of ladder up to top of building given by

$\qquad AB = \dfrac{7.5}{\sin 75} = 7.76(\ldots)$.

Ladder length above landing place given by

$\qquad BC = \dfrac{1.074}{\sin 75} = 1.11(\ldots)$.

Total ladder minimum

$\qquad = AB + BC = 8.88$ m.

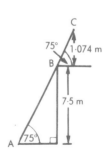

A15

Original surface area given by $\pi r l = \pi \times 5 \times 13 = 204.(\ldots)$. Top cut off surface area given by $\pi \times 2.5 \times 6.5 = 51.(\ldots)$. Hence painted area $= 153$ cm^2.

A16

(i) $C = \sqrt{(7^2 + 24^2)} = 25$

(ii)   The table is filled in as:

a	b	c
3	4	5
5	12	13
7	24	25
9	40	41
11	60	61
13	84	85
15	112	113
17	144	145

The c is always one more than the b. The b numbers have differences of 4, 8, 12, 16. . . .

A17

The four quarter circles give the same perimeter as a full circle of radius 5 m, which gives a perimeter of $\pi \times 2 \times 5 = 31$ cm.

A18

(a)   1.18 cm.

(b) $\frac{6}{2}(0.42 + 1.18) = 4.8$ cm^2.

(c)   (i) $4.8 \times 3 = 14.4$ cm^2;   (ii) $14\,400\,000$ cm^3.

GRADE CHECKLIST

FOR A GRADE F

You should be able to:

- calculate the perimeter and area of a rectangle, triangle and a plane figure made up of rectangles;
- calculate the volume of a rectangular block or of a solid made up of cubes;
- calculate the circumference of a circle with a suitable value of π;
- find the area of any plane figure by counting squares.

FOR A GRADE C

You should also be able to:

- calculate the area of a parallelogram and a circle;
- calculate volume and area of a cylinder;
- use the rule of Pythagoras to calculate sides in a right-angled triangle;
- use sine, cosine and tangent to calculate sides and angles of a right-angled triangle.

FOR A GRADE A

You should also be able to:

- calculate the area of a trapezium;
- calculate volumes of prisms, spheres and cones;
- calculate the surface area of spheres and cones;
- carry out calculations involving the use of right-angled triangles in three-dimensional situations.

STUDENT'S ANSWER - EXAMINER'S COMMENTS

QUESTION

The circle shown has centre 0 and radius 6 cm.
PB has length 8 cm.

Calculate (a) the size of the angle PAB
        (b) the length of AP.

66 **diagram not consistent with the working out for part (a)** 99

66 **not said** *why* **it is a right angle** 99

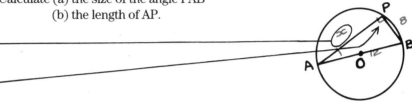

Answer (a)   not to scale

$$\frac{OPP}{HYP} = SIN$$

66 **Poor notation.**
sin x = 0.666
x =
should have been used 99

$$SIN\ X = \frac{8}{12} = 0.666 \quad (REC.)$$
$$SIN\ OF\ 0.666\ (REC)$$
$$= 41.8$$

(b)

66 **correct answer** 99

Answer (a) $41.8°$

Answer (b) $8.9\ CM$

66 **Pythagoras would have been safer since you use the given information** 99

$$COS\ 41.8 = \frac{x}{12}$$

66 **a correct method but this line should read** **12 X cos 41.8 = 8.9** 99

$$41.8 \times 12 = x$$
$$x = 8.9$$

DRAWING

GETTING STARTED

All GCSE syllabuses include drawing, in particular geometrical type drawing. It is the intention of all GCSE examinations to test the use of geometry in conveying information and solving problems, often by means of drawing and measuring. So you do need to be able to draw certain shapes, and accurately!

You need to be able to use a **protractor** (sometimes called an angle measurer) to measure and draw acute, obtuse and even reflex angles. You must be able to use a pair of **compasses** to draw a circle or to measure a given distance. It is important that you can use a **set square** both to draw a right angle and to draw parallel lines. Of course you also need to be able to draw accurate lines and measure with a ruler. Be careful to use the *centre* of the marks on the ruler when you are using it to measure, or else you could be inaccurate.

When asked to draw or to construct a diagram, a common accuracy that is looked for is to be no more than 1 mm out on lengths of lines, and no more than 1 or 2 degrees out on angles. So be warned; be accurate, or you will certainly lose marks.

Only when you are confident that you can use these items of equipment can you be confident in your ability to draw accurately when necessary.

TRIANGLES
RECTANGLES
A QUADRILATERAL
CONSTRUCTIONS
LOCUS
BEARINGS

USEFUL DEFINITIONS

Right angle	an angle that measures 90°
Acute angle	one that is less than 90°
Obtuse angle	one that is between 90° and 180°
Reflex angle	one that is between 180° and 360°
∠	shorthand for angle
Quadrilateral	a shape with only four straight sides
Trapezium	a quadrilateral with a pair of opposite sides parallel
Parallelogram	a quadrilateral with both pairs of opposite sides parallel
Polygon	a flat shape with many straight sides
Pentagon	a polygon with five sides
Hexagon	a polygon with six sides
Octagon	a polygon with eight sides
Vertex	the 'sharp bit' of an angle
Perpendicular	at right angles to

E S S E N T I A L   P R I N C I P L E S

1 > TRIANGLES

You can be given certain information about a *triangle* and be expected to draw it accurately.

ALL ANGLES KNOWN AND A SIDE GIVEN

You should already know that the three angles of a triangle add up to 180°. So if you are told two of the angles the other one is easily worked out. If one angle is 40° and another 60°, then the third angle must be 80° (i.e. 180 − (40 + 60)). It is usually best to try and draw the triangle so that the side *given* is the *base* of what you are to draw and then to draw lines at the angles specified.

WORKED EXAMPLE 1

Draw a triangle *ABC* where ∠A = 60°, ∠B = 40° and *AB* = 7 cm.

It is helpful to *sketch* the triangle *ABC* first. This will help you to see exactly what you need to draw.

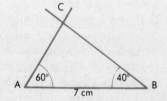

   Start by drawing the 7 cm line with a ruler. This is the side *given* and you should draw it as the *base* of the triangle. Then use a *protractor* to draw a faint line at 60° from the left-hand end of the base line, and a faint line at 40° from the right-hand end. Draw these faint lines so that they cross over. When you know *where* they cross you can draw the lines more heavily up to that point.

TWO SIDES AND THE INCLUDED ANGLE

This is where you are told one angle and the length of each side next to the angle. It is common to draw the *longest side* as the *base*. Then draw a line at one end of the base at the required angle and to the required length. Now join up to complete the triangle.

WORKED EXAMPLE 2

Draw a triangle *ABC* where ∠A = 50°, *AB* = 4 cm and *AC* = 6 cm.

Start by drawing the *longest side* as the *base* of the triangle (bottom side). Draw this as accurately as possible with a ruler, then use a *protractor* to draw a faint line at 50° at the left-hand end. Measure accurately 4 cm up this line and draw more heavily. You can then join this end to the right-hand end of the 6 cm base line.

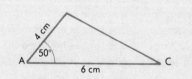

ALL THREE SIDES GIVEN (NO ANGLES)

Again you would usually start by drawing the *longest side* as the *base* of the triangle. Then use a pair of *compasses* to 'arc' each of the other two lengths.

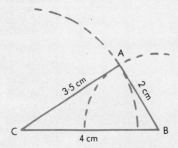

<table>
<tr><td>

</td><td>

Draw a triangle *ABC* where $AB = 2$ cm, $BC = 4$ cm and $AC = 3.5$ cm.

Start by drawing the base length of 4 cm as accurately as possible with your ruler. Now, you need your pair of compasses. Make the distance between the sharp end and the pencil end 3.5 cm.

</td></tr>
</table>

Then, with the sharp end positioned at the left-hand end of the base line, draw a faint *quarter circle* above the base line (as shown by the dotted line) – we call this 'arcing'. Repeat this for the distance of 2 cm from the other end. Where these two arcs *cross* is the point to which you draw the other two lines, giving you all three sides of the triangle.

EXERCISE 1

Draw these as accurately as you can, so use a sharp pencil.

Draw the triangle *ABC* where
(i) $\angle A = 70°$, $\angle B = 30°$, $BC = 6$ cm.
(ii) $\angle B = 60°$, $AB = 4.7$ cm, $BC = 5.6$ cm.
(iii) $AB = 4$ cm, $BC = 7$ cm, $AC = 5$ cm.

2 ▷ RECTANGLES

To construct a *rectangle*, all you need to be told is its *length* and *breadth*. For example, to draw a rectangle that measures 8 cm by 4 cm you would start by drawing the *base length* of 8 cm. Then faintly draw the angles of 90° at each end, either by *set square* or *protractor*. Measure 4 cm up each line, then join the tops to give you the rectangle.

3 ▷ A QUADRILATERAL

You could be asked to construct a *quadrilateral* to some particular size. You would be given sufficient information to allow you to start with a *base* line and angles on either side (like drawing a triangle with two angles and a side given). The length of at least one other side would be given. This can be drawn in *after* faint lines have been drawn from each end of the base line at the appropriate angles.

EXERCISE 2

Construct the quadrilateral *ABCD* where $AB = 4$ cm, $AD = 10$ cm, $CD = 7$ cm, $\angle A = 70°$, $\angle D = 80°$. Measure the length of *BC*.

4 ▷ CONSTRUCTIONS

When asked to 'construct', you need to show *how* you have constructed the line or shape, otherwise the examiner will assume that you have guessed or used a protractor and you will lose marks.

The constructions given here are the ones expected to be used in the examination. Of course other suitable alternative 'constructions' will always gain full marks also.

LINE BISECTOR

The diagram illustrates how to cut the line *AB* *exactly in half* (in other words to *bisect* it). Set your compasses to about three-quarters of the length of the line *AB*, and with the sharp point at one end draw a faint semi-circle; then repeat from the other end keeping the compass arc the same. You need to find out where the two semicircles cross over. The straight line between these points will give the *line bisector*.

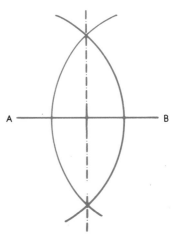

ANGLE BISECTOR

The diagram illustrates how to bisect the angle *PQR*. With your compasses set at about 2 cm, put the sharp end at the vertex of the angle (*Q*), and arc the angle as shown with a faint line. At both points where this arc cuts the sides of the angle, and using the sharp end of the compasses, draw another faint arc across the angle as shown. Where these *cross*, join to the vertex at *Q*. The resulting line is the line that bisects the angle (cuts it in half).

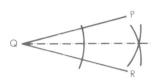

A RIGHT ANGLE

Often associated with a 'perpendicular line', a right angle can always be drawn with a set square, or more accurately with ruler and compasses, as shown here.

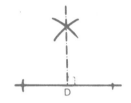

To construct a right angle at the point *D* on the line shown, you need two marks at *equal distances* on either side of the point (the line may well need extending to do this).

The marks are made usually by compasses and simply arced through at either side. Then repeat what you would normally do to bisect a line but this time use the *two arcs* you've just made for the sharp point of your compasses. You don't need to draw full semi-circles as we are only interested in *one side* of the line – that side on which we want to construct the right angle. Find where the arcs you have drawn cross over. Join up with the point *D* and you have your right angle.

A 60° ANGLE

To construct an angle of 60° at point *E* on the line *EF* shown, you can set your compasses to any distance you like. Then with the sharp end at point *E* draw faintly the quarter circle that arcs through *EF*. Where this has cut *EF*, put the sharp end of the compasses (keep it the same distance again!), and draw the arc faintly that goes from *E* and cuts through the first arc. Where these *cross* you can draw a straight line to point *E* and you have your angle of 60°.

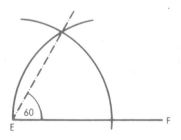

OTHER ANGLES

You can now construct many more angles. If you bisect a right angle you can make 45°, and you can bisect again to make $22\frac{1}{2}°$; or you can bisect 60° to make 30°, and bisect again to make 15°. If you really wish to (it is fun to try!), you can put these together to create angles like 75°, (45 + 30) or (60 + 15).

EXERCISE 3

Construct angles of (i) 30° and (ii) 75°.

A PERPENDICULAR LINE FROM A POINT

In the diagram shown here, there is a line AB. We are going to construct the *perpendicular line* from point P to this line. Use your compasses to draw as wide an arc as you can cutting AB at two points, with P as the centre of the arc. Then from *each point on AB that you have arced*, and using the same compass opening, make an arc *under* the line AB as shown. The two arcs will cross, giving you the point from which to draw the perpendicular line from P to the line AB.

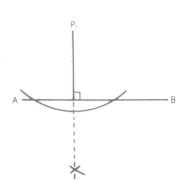

---

5 ➤   LOCUS

A *locus* is the collection of all possible points relating to some simple (!) rule. For example, the locus of all the points 3 cm away from a dot will be a circle of radius 3 cm with that dot as its centre. Generally, to find a locus we can fix in some points on a drawing and *go on fixing points* until a pattern emerges.

WORKED EXAMPLE 4

Two pegs are fixed in the ground a distance of 5 metres apart. A rope is tied to each peg leaving 6 metres of rope loose. A peg is put into the ground at a point P so that this 6 metres of rope is taut, as shown in diagram (a). Sketch (using 1 cm to 1 m) the locus of this point P.

With compasses we can find various positions of P as shown in diagram (b). Each point was found by finding the *vertex* of a triangle which has its base between the two pegs A and B, and where the top two sides add up to 6 cm. We can now see a pattern and could draw an *oval shape* going through each point P. This is then the *locus* of the point P when we've drawn that shape in.

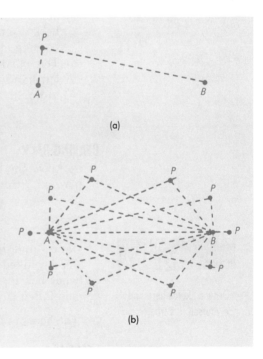

(a)

(b)

---

6 ➤   BEARINGS

If you have ever been out walking on the hilltops when the mist has come down you will realise how important it is to be able to read a *compass*, as illustrated here. On a compass the magnetic needle moves round to point towards the North Pole (not to the north on your compass). The compass then needs to be rotated so that the needle *does* point towards the 'N' on the compass. Then, by moving your map around so that it too is pointing to the north (there should be an arrow on the map indicating north), you can tell which way you should be walking, or sailing or even flying!

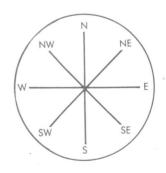

Around your compass you may well see other numbers like 005, 080, 260. These are 3-figure bearings and represent the angle from north which that direction is making. You should see that east will be 090, south-east is 135, south is 180, west is 270 and north-west is 315°. You ought to be able to put any of the eight main compass points into a 3-figure bearing – see if you can.

| **WORKED EXAMPLE 5** | |

If you found yourself right in the middle of Whitwell wood where all eight paths meet:
(a) which path would you take to get to Firbeck Common? What is its bearing?
(b) where do you come to if you take the north-west path?

(a) From the centre of the wood, draw or imagine a north line. Put your protractor on this to measure the angle between north and the path to Firbeck Common – you should read about 60°. Hence the path to Firbeck Common is the one on a bearing of 060°.
(b) The north-west path is the top left-hand corner one that leads to Bondhay Farm.

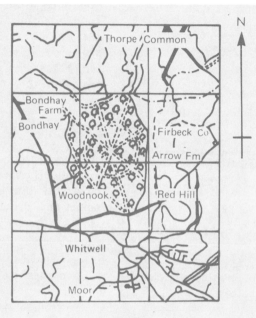

BEARING BACK

If we know the bearing of a point B from a point A, then we can always find the bearing of A from B by adding on 180°.

| **WORKED EXAMPLE 6** | |

" Draw a sketch here and convince yourself it is true. "

James and John were out on a hill walk when James fell into a pothole! John walked on a bearing of 075° to find help. When he found help, on what bearing should he walk back in order to find the pothole that James had fallen down?

The bearing back again will be 180 + 75, which is 255°.

EXERCISE 4

Swinton is on a bearing of 160° from Wath. What is the bearing of Wath from Swinton. If Swinton is 1 km east of Wath, how far north is Wath from Swinton?

S O L U T I O N S   T O   E X E R C I S E S

S2

If you started with AD as the *base*, then drew AB at 70° and DC at 80°, you should have ended up with BC measuring 8 cm. However, if you are no more than 2 mm out, you have done well.

S3

(i)   You should have constructed 60° then bisected it.

(ii) The simplest way would have been to construct 60°, then another angle of 60° on the first (to give 120°). Bisect, then bisect the angle again, to leave a 15° on top of a 60°, giving 75°.

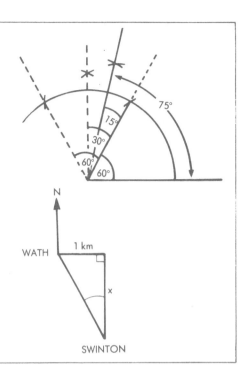

S4

The return bearing is 180 + 160, which is 340°. A diagram to help us with the next part is shown here. The angle marked is 20°, since 360 − 340 is 20°. You can either draw this accurately and measure the length x, or use trigonometry. Either way you should get an answer of 2.7 km.

EXAMINATION TYPE QUESTIONS

Q1

A little old lady, we'll call her Gran, read in a book that trees were dangerous if they were over 30 ft high, and she was sure that the tree in her garden was over 30 ft high.

Gran was about 5 ft high and when she stood 20 ft away from the tree on level ground she used a 'clinometer' to show her the top of the tree was 48° to the horizontal, as shown in the sketch.

(a)   Use an accurate scale drawing of 1 cm to 5 feet to find the height of the tree.

(b)   Would Gran say it was dangerous?

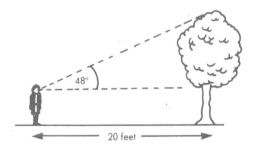

Q2

One night a smuggler set sail from France in a north-westerly direction. At the same time, due east along the French coastline at Cherbourg, a French customs patrol boat set sail on a bearing of 060°, and after sailing for 4 kilometres caught the smuggler 'red handed'. How far along the coastline from Cherbourg had the smuggler set sail? (Draw an accurate scale sketch of the situation using 1 cm to 1 km.)

Q3

(a) You are facing due west and you turn clockwise to face north-east. Through how many degrees do you turn?

(b) State two directions which are at right angles to south-west.

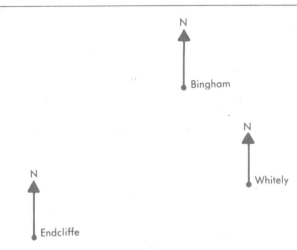

Q4

From the sketch map, with a scale 1 cm to 2 km:
(a) find the bearing of Bingham from Endcliffe
(b) find the distance from Endcliffe to Whitely.

(NEA)

Q5

A plane left an airport on a bearing of 060° flying a distance of 120 kilometres, landing at an airport to pick up vital medical supplies. The plane then flew for 250 kilometres on a bearing of 150°. What single journey must the plane now fly to return to its starting point?

Q6

A 'Mayday' call is heard by the three boats marked on this diagram by dots – H, M and S. They can tell by the reception that the mayday call is nearer to ship M than it is to ship H, and the call is less than 3 km from ship S.

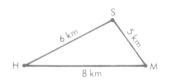

Draw an accurate plan of the position of the ships (using 1 cm to 1 km) and indicate all the possible positions of the mayday call.

Q7

Complete the plan of a model, shown opposite, by putting in a fold mark with a dotted line that bisects the angle at A.

Q8

Draw a rectangle with the same area as this trapezium. Explain your answer.

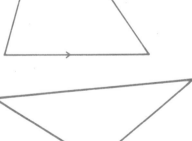

Q9

Calculate as accurately as you can the area of this triangle. Explain how you arrive at your answer.

Q10

An inscribed circle is one that is drawn inside a triangle where each side is a tangent to the circle. The centre of this circle is where the angle bisectors all intersect. In the triangle given, draw an inscribed circle.

Q11

Draw a circle radius 4 cm and construct a tangent to the circle at any point.

OUTLINE ANSWERS TO EXAM QUESTIONS

A1

You can draw this picture quite accurately, using single lines for Gran and the tree, and you will then be able to measure the height of the tree as 27 ft (don't forget Gran's height in this). Hence Gran would say that her tree is not dangerous.

A2

You can sketch this picture of the situation. Then, since the smuggler sets sail due east of the customs, angle x will be $90 - 60$ which is $30°$, and hence angle y will be $180 - (30 + 45)$ which is $105°$.

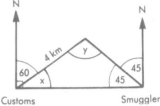

You can now draw this out accurately, starting with the 4 km line, and by doing this you will be able to measure the distance apart when both ships set sail. This will be $5\frac{1}{2}$ cm, which will mean that the smuggler set sail $5\frac{1}{2}$ km along the coast from Cherbourg.

A3

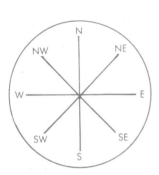

(a) If you look at the diagram, or a compass, you will see that from due west clockwise you go through NW and N to get to NE, which is three lots of $45°$, hence $135°$.

(b) Again, look at the diagram, or a compass, and $90°$ will be two compass points round either way. So you will go from SW either through S to SE, or through W onto NW.

A4

(a) Draw a line from Endcliffe to Bingham, then measure the angle made from the north line at Endcliffe. This will be $45°$, hence the bearing will be either $045°$ or north-east.

(b) Measuring with a ruler the distance on the map, you will get 6 cm, and since the scale of the map is 1 cm to 2 km this will represent 12 km.

A5

If you draw this out to a suitable scale, you will end up with a drawing as shown. The dotted line indicates the return journey which will be approximately 270 km on a bearing of $306°$.

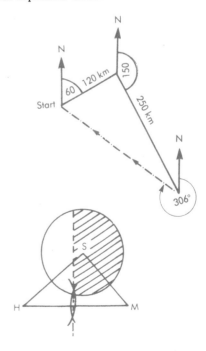

A6

You can draw a triangle to the scale suggested and get a triangle the same shape (but larger) than the one shown here. If the call is nearer to M than H, then we need a bisector drawn between H and M, as we have done before, and the call is shown to be to the right of this line. As the call is less than 3 km from S, there will be a circle around S of radius 3 km in which the mayday call will lie. Hence we obtain the shaded part which is both nearer M than H and less than 3 km from S.

A7

You need to construct an angle bisect line all the way along your diagram.

A8

One way is simply to measure the dimensions, calculate the areas and then draw a suitable rectangle! Another way is to find the halfway mark on each sloping side, and use this to mark the sides of the rectangle as in the diagram. Either way is acceptable as long as it is carefully explained.

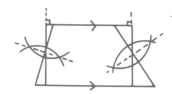

A9

The area is found by $\frac{1}{2} \times$ base $\times$ perpendicular height. So choose any base, then construct the perpendicular line from the vertex to the base, and hence measure accurately the base length and the height. Multiply out and you should get an answer close to 5.1 cm^2.

A10

Bisect each angle carefully, and although only two bisectors are necessary, the third one acts as a check on the other two. Where they cross is the centre, then simply draw the circle.

A11

Draw the circle, choose a point on the circumference, then draw in, faintly, the radius to this point. Construct a line perpendicular to the radius at this point which can be extended to give the tangent.

GRADE CHECKLIST

FOR A GRADE F

You should know:
 the eight main compass directions;
you should understand:
 what a 3-digit bearing is;
and be able to:
 draw simple plane figures to given dimensions;
 construct scale drawings, and use them to gain information;
 use the eight main compass directions together with 3-digit bearing.

FOR A GRADE C

You should also be able to:
 make accurate drawings of plane figures to given measurements;
 use accurate drawings to solve problems;
 identify and use simple loci.

FOR A GRADE A

You should also be able to:
 construct a perpendicular line from any point to a given line.

STUDENT'S ANSWER - EXAMINER'S COMMENTS

QUESTION

(i) In the rectangle below draw accurately the sets
$P = \{X: XA = XB\}$ and $Q = \{Y: YA = 4.5\text{ cm}\}$

(ii) Given that $P \cap Q = \{H, K\}$, draw the circle that passes through B, H and K.

(iii) Find the area of triangle *BHK* in two different ways, stating clearly the methods you have used.

Comment on the degrees of accuracy of your answers.

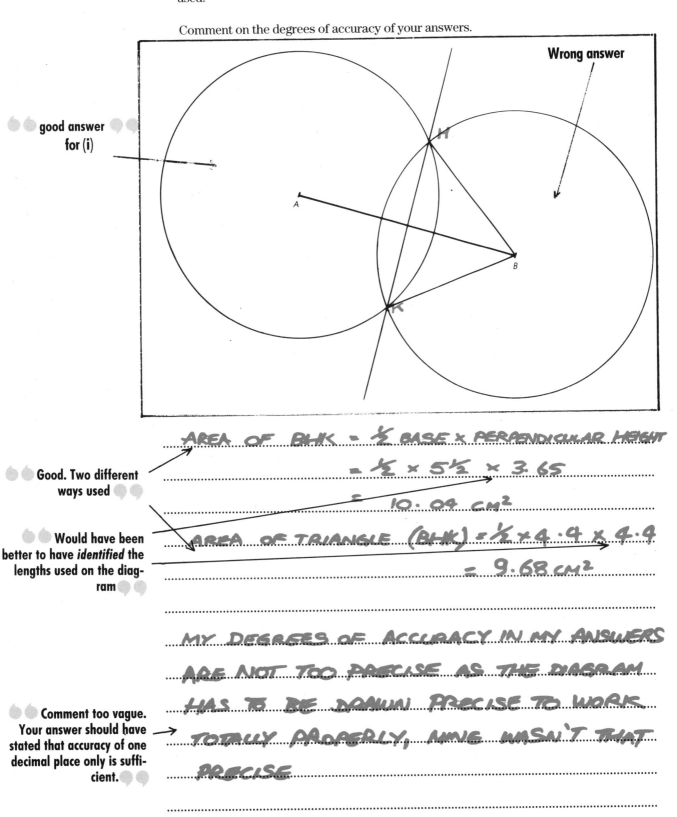

Wrong answer

good answer
for (i)

Good. Two different
ways used

Would have been
better to have *identified* the
lengths used on the diagram

AREA OF BHK = ½ BASE × PERPENDICULAR HEIGHT

= ½ × 5½ × 3.65

= 10.09 CM²

AREA OF TRIANGLE (BHK) = ½ × 4.4 × 4.4

= 9.68 CM²

Comment too vague.
Your answer should have
stated that accuracy of one
decimal place only is sufficient.

MY DEGREES OF ACCURACY IN MY ANSWERS
ARE NOT TOO PRECISE AS THE DIAGRAM
HAS TO BE DRAWN PRECISE TO WORK
TOTALLY PROPERLY, MINE WASN'T THAT
PRECISE

TRANSFORMATION
GEOMETRY

You will need to look carefully at exactly which parts of this chapter are in *your* syllabus. You will also need to consider the *level* of the examination you are sitting. Only the *higher* levels will include *matrix transformations*, and its related parts.

The essence of *transformation geometry* is looking at how shapes change position and size according to certain rules. You will not be required to prove anything, but you will need to *describe* how a shape has changed, or to *actually change* the shape itself to the rules given.

There are some definite links in this chapter with the ideas of *symmetry* you met in Chapter 7. This connection is often used in actual examination questions.

USEFUL DEFINITIONS

Transformation	a change of position
Reflection	a mirror image the other side of a line
Rotation	a turn around some fixed point
Translation	a slide *without* any turning
Enlargement	a change in size (in mathematics an enlargement can be smaller!) to create a similar shape
Matrix	a collection of numbers in some specific order
Vector	a single list in a particular order
Tessellation	a regular pattern created from shapes that leave no spaces at all

REFLECTIONS
ROTATIONS
TRANSLATIONS
ENLARGEMENTS
SHEAR
COMBINATION OF
    TRANSFORMATIONS
TESSELLATIONS
MATRICES
VECTORS
MATRICES WITH
    TRANSFORMATIONS

E S S E N T I A L   P R I N C I P L E S

When you are asked to *reflect* a shape in a given line, then there are two ways of doing it. For example, if you were asked to reflect the rectangle in the line *AB* shown here you could trace it, then flip the tracing paper over so that the reflection appears under the line with the tracing of line *AB* exactly on top of the original (especially *A* on top of *A* and *B* on top of *B*). Then, with a pencil, press down on each corner of the tracing to make a 'dint' in the paper underneath. Take the tracing away, join up the dots and this will give you the reflection. If you do this carefully it is the easiest way.

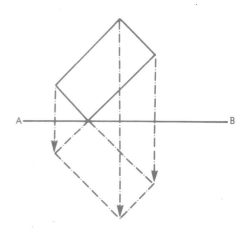

The other way is to draw faint lines from each vertex of the rectangle perpendicular to the line *AB*, then measure the distance from each point to the line, and its reflection is exactly the same distance away from *AB* the other side of *AB* along this faint perpendicular line you've just drawn. Put a dot at this position. Do this for each vertex then join up the dots. On squared paper particularly this method is the most accurate.

When you have completed a reflection, by either method, you can check it by making sure that the line you reflected on (sometimes called the mirror line) is a line of symmetry for your drawing, and hence the reflected shape will have all its dimensions and angles the same as before – it's just the position that is different.

You need to be able to *rotate* any given shape through 90° (either way) and through 180°. For example, the illustration here shows the triangle *ABC* has been rotated about the centre of rotation ∗, through 90° clockwise ($\frac{1}{4}$ turn) to position *A'B'C'*, then another 90° clockwise, which is equivalent to a rotation of 180° to *A"B"C"*.

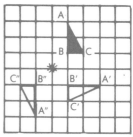

One way to do the 90° rotation is again to use tracing paper and trace the shape you are to rotate with the centre of rotation marked with a suitable + (following the lines of the grid). Then with your pencil point on the centre of rotation, turn the tracing paper until you can see the + back on top of itself having turned 90°. Now you can press on each vertex of the triangle, take off the tracing paper and join up your dots. Remember that if you are working on squared paper the shape will end up with the vertices on the corners of the squares and that each line of the new triangle will have turned through 90° also. The resultant shape should be the same size, have the same angles, but just be in a different position. To do the 180° rotation you would DO the same, but of course turn through 180°.

The other way to rotate 90° is to, again, look at the position of the vertex from the centre of rotation and, for example, if *A* is 3 up and 1 to the right, then it rotates to 3 to the right and 1 down from the centre of rotation. This works for all points, and then join them up. To rotate the 180°, point *A* this time will change to 3 down and 1 to the left, and so on for the other points. When you've practised and mastered it, this last technique is by far the easiest and quickest!

3 ▷ TRANSLATIONS

A *translation* is a movement along the plane without any rotating, reflecting or enlarging. It is described by a movement horizontally and a movement vertically which we put together as a *vector*.

For example, the heavily shaded triangle here has been translated to position A by moving 2 to the right and 1 up. (Notice that every point on the triangle has moved this way.) We write this movement as a vector $\begin{pmatrix} 2 \\ 1 \end{pmatrix}$.

In a similar way we write the following translations – notice how and when we use a negative sign. Each one is from the heavily shaded triangle.

To B we move 3 to the right and 2 down $= \begin{pmatrix} 3 \\ -2 \end{pmatrix}$.

To C we move 3 to the left and 2 down $= \begin{pmatrix} -3 \\ -2 \end{pmatrix}$.

To D we move 2 to the left only $= \begin{pmatrix} -2 \\ 0 \end{pmatrix}$.

To E we move 3 to the left and 1 up $= \begin{pmatrix} -3 \\ 1 \end{pmatrix}$.

4 ▷ ENLARGEMENTS

The idea of an *enlargement* is to make a shape larger and in a specific place. Hence you will be given a *scale factor* which tells you how many times bigger each line will be, and a centre of enlargement which will determine where the enlargement ends up! For example, to enlarge the shaded square in diagram (a) with a scale factor of 2 from the centre of enlargement ∗, the distance from the centre of enlargement to each vertex in the shape is multiplied by 2 (doubled).

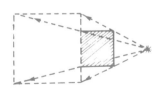

This is shown on diagram (a) by the faint dotted lines. The enlarged shape will have all its dimensions multiplied by the scale factor, but keep all its angles the same. In other words it is a 'similar' shape.

Enlargements will tend to be more accurate if done on squared paper when you can use the lines instead of measuring. For example, in diagram (b) we see a square being enlarged from ∗ with a scale factor of 3.

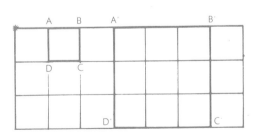

Since A is one square along from ∗, then its new position A' will be $1 \times 3 = 3$ squares along. Since C is 2 along and 1 down, its new position C' will be $2 \times 3 = 6$ along and $1 \times 3 = 3$ down.

You can use either way for any positive scale factor, even $1\frac{1}{2}$, when you simply multiply distances from the centre of enlargement by $1\frac{1}{2}$. However, if you are asked to do an enlargement by $\frac{1}{2}$, then the procedure is the same, only this time you will end up with a shape only half its size. For example, if, in diagram (b), we were asked to enlarge the shape $A'B'C'D'$ from centre of enlargement $*$ with a scale factor of $\frac{1}{3}$, then the result would be the drawing of square $ABCD$ as shown.

NEGATIVE ENLARGEMENT

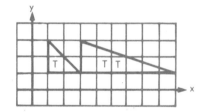

> This is the most difficult one, and is often done badly. So go carefully.

You have seen how to transform a shape with a positive enlargement. However, you could be given a *negative* scale factor. In that case we draw the lines back through the centre of enlargement and then enlarge as shown here in an enlargement of the small triangle through $*$ with an enlargement of -2. Notice how, with a negative scale factor, the enlarged shape has ended up 'upside down' but still a similar shape.

ONE-WAY STRETCH

This is when we enlarge in one direction only. For example, in the diagram given here, the triangle T has undergone a *one-way stretch* of scale factor 3 with the y-axis *invariant* leaving the stretched shape TT. The y-axis being invariant means that anything on the y-axis stays where it is and the shape gets stretched away from that line. The distance from the invariant line to any point is multiplied by the scale factor to find its stretched distance from the invariant line.

You are only likely to be asked to transform with a one-way stretch with either the x- or the y-axis as the invariant line.

5 >   SHEAR

A *shear* also has an invariant line, generally either the x- or the y-axis. It has a 'shear factor' which indicates how much the shape is to be 'pushed' over:

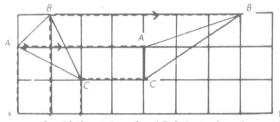

For example, if the triangle ABC is to be sheared with a 'shear factor 2' and the line XY is invariant, then each point is to be moved parallel to the invariant line in the positive sense (i.e. to the right with x-axis invariant, or up if the y-axis is invariant). The distance moved is:

(perpendicular distance from the invariant line) × (shear factor).

Hence A is to move $2 \times 2 = 4$ units
$\qquad$ B is to move $3 \times 2 = 6$ units
$\qquad$ C is to move $1 \times 2 = 2$ units.

This moves the shape to $A'B'C'$ as seen in the diagram. You now know why we used the word 'pushed' over.

EXERCISE 1

On a grid with x from -5 to 8, and y from -6 to 9 draw the triangle with vertices (corners) $A(1, 1)$, $B(1, 3)$, $C(2, 1)$. On this triangle 'do' the following transformations:

(i)   reflection in the x-axis ($y = 0$);

(ii)   rotation, around $(0, 0)$, of $90°$ anticlockwise;

(iii)   translation of $\begin{pmatrix} 3 \\ -4 \end{pmatrix}$;

(iv)   enlargement scale factor 3 from centre of enlargement $(3, 0)$;

(v)   enlargement scale factor -2 from centre of enlargement $(0, 0)$;

(vi)   a one-way stretch of scale factor 3 with x-axis invariant;

(vii)   a shear with shear factor 2, x-axis invariant.

(None of your answers should be overlapping another!)

---

6 > COMBINATION OF TRANSFORMATIONS

If we *combine* two transformations, say a rotation of $90°$ clockwise around the origin and a reflection in the y-axis, then the *order* in which we do this will make a difference. For example, this rotation and enlargement have been combined to give the following situations on the same triangle.

Figure (a) shows the rotation of shaded triangle to A followed by the reflection to give B. Figure (b) shows the reflection to A followed by the rotation to give B. The same transformations, when combined in a different way, give different results, hence the *order* in which you combine transformations is important. However, this is not always so – see if you can find some combinations that make no difference to what is done first.

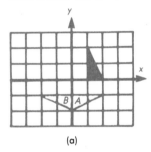

(a)

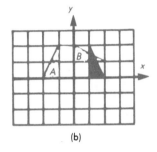

(b)

---

7 > TESSELLATIONS

Not really a transformation at all, but a pattern. A *tessellation* is a regular pattern with *one* shape that could cover a large area without leaving gaps (except at the very edge). Below are some examples of tessellations.

Each tessellation is made from one plane shape and could continue its pattern to fill in a large area leaving no gaps. It is true to say (and you can test it out for yourself) that every triangle and every quadrilateral will tessellate!

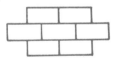

---

8 > MATRICES

A *matrix* is a collection of information put in tabular form in a specific order. For example, in the County Cricket table illustrated here, the *numbers* form the matrix. They are in specific order as indicated by the labelling at the top and up the side.

COUNTY TABLE

	P	W	L	D	B⁺	B	Pts
Gloucs 3	17	9	1	7	36	50	230
Essex 4	15	6	4	5	33	4⁷	1⁷6
Han'ts 2	15	5	4	6	36	48	164
Surrey 6	16	5	6	5	34	50	164
Yorks 11	17	4	3	10	48	42	162
Worcs 5	17	4	5	8	42	50	156
Notts 8	15	4	2	9	40	50	154
Leics 6	16	4	4	8	39	47	150
Kent 9	15	4	4	7	28	49	141
Derbys 12	15	4	3	8	24	50	138
Nor'ts 10	15	4	2	9	36	36	136
Som'set 17	15	3	2	10	40	31	119
Lancs 14	16	3	2	11	34	36	118
W'wick 15	16	2	3	11	34	43	109
Sussex 7	16	2	7	7	28	37	97
Mid'sex 1	16	1	8	7	29	49	94
Glam 12	16	1	5	10	29	34	79

*1985 positions in brackets.
*Yorkshire total includes eight points for drawn match where scores finished level.

ORDER OF A MATRIX

The *order* of a matrix is given by an ordered pair of numbers $a \times b$ where a is the number of *rows* in the matrix and b is the number of *columns*. So, for the County Cricket table above the order is 17 by 7.

ADDITION AND SUBTRACTION

Since the information in the matrix is in a specific order we can add matrices together (or subtract where necessary) as long as it is sensible to do so; in other words, only when the information given in the matrices is about the same thing, and when the matrices are of the same order. We add (or subtract) the *corresponding numbers* in each matrix.

WORKED EXAMPLE 1

At the beginning of the last day of the British Games the top of the medals table was as shown here:

	G.	S.	B.	Total
England....	23	14	18	55
Wales........	19	21	16	56
N. Ireland	16	21	18	55
Scotland....	16	18	22	56

The medals won during the last day were:

	G.	S.	B.	Total
England....	4	3	2	9
Wales........	3	3	4	10
N. Ireland	4	2	3	9
Scotland....	2	5	4	11

Add together these matrices to find the final medal table of the British Games that year.

Since the information given in each matrix is of the same type and since they are both in the same order, then we may *add* the numbers together to give a final medals table as:

	G.	S.	B.	Total
England....	27	17	20	64
Wales........	22	24	20	66
N. Ireland	20	23	21	64
Scotland....	18	23	26	67

WORKED EXAMPLE 2

Before a stocktaking check the following items should have been in stock at a shop:

	Small	Medium	Large
V-neck	8	15	4
Polo	9	12	5
Crew neck	6	18	8

The stocktaking check showed the following:

	Small	Medium	Large
V-neck	7	13	4
Polo	7	9	4
Crew neck	6	15	6

Write down as clearly as possible what the shop has missing.

Since the matrices are both about the same thing and are of the same order, we may *subtract* them to find the difference. Doing this we obtain the matrix shown, which is the clearest way of giving the information.

	Small	Medium	Large
V-neck	1	2	0
Polo	2	3	1
Crew neck	0	3	2

ROUTE MATRICES

ONE STAGE

Given a *network* of paths as, say, in the diagram, we can describe this by using a *matrix* to tell us how many direct *one-stage* routes there are from one point to the other (e.g. A to B without going through any other points).

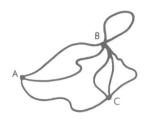

Hence the matrix to describe this network of routes will be as written here and is read from left to right.

		To		
		A	B	C
	A	0	2	1
From	B	2	2	3
	C	1	3	0

Note that from B to B we see there are two routes. This is because we can go either way along the path shown.

TWO STAGE

A *two-stage* route is one that combines two 'one-stage' routes to get from one point to another.

WORKED EXAMPLE 3

Find the two-stage route matrix for the diagram.

This is found by considering each pair of points.
Hence A to A has five different routes:
(1) A to A clockwise and again clockwise to A.
(2) A to A clockwise then again anticlockwise to A.
(3) A to A anticlockwise and again anticlockwise to A.
(4) A to A anticlockwise then again clockwise to A.
(5) A to B then B to A.

Also, A to B has two different routes. Try to find them.

Then A to C has two routes:
(1) A to B then B to C on right-hand road.
(2) A to B then B to C on left-hand road.

When you have gone through each pair of points you will end up with the two-stage route matrix

		To		
		A	B	C
	A	5	2	2
From	B	2	5	0
	C	2	0	4

SQUARING

The other way to find the two-stage route matrix is to multiply the matrix by itself (to square it).

> **WORKED EXAMPLE 4**

Find the two-stage route matrix from the above diagram by squaring it.

First we find the one-stage route matrix, i.e.

$$\begin{pmatrix} 2 & 1 & 0 \\ 1 & 0 & 2 \\ 0 & 2 & 0 \end{pmatrix}$$ (note the use of the brackets and no other labelling)

You now multiply $\begin{pmatrix} 2 & 1 & 0 \\ 1 & 0 & 2 \\ 0 & 2 & 0 \end{pmatrix}\begin{pmatrix} 2 & 1 & 0 \\ 1 & 0 & 2 \\ 0 & 2 & 0 \end{pmatrix}$ by combining *each row* in the first

matrix with *each column* in the second; by multiplying the first number in the row by the first in the column, then the second in the row by the second in the column, and the third in the row by the third in the column, then adding to give a single number. Each row in the first combines with each column in the second to give a single number in the answer matrix.

So this will given an answer of

$$\begin{pmatrix} 5 & 2 & 2 \\ 2 & 5 & 0 \\ 2 & 0 & 4 \end{pmatrix}$$

which is the same answer as before.

MATRIX MULTIPLICATION

This way of combining matrices can be done between any two matrices provided the second matrix has the same number of rows as the number of columns in the first.

You will find that a $(x \times y)$ matrix, multiplied to a $(y \times w)$ matrix, gives an answer of order $(x \times w)$.

EXERCISE 2

(i)   Write down the one-stage route matrix from the network shown in the diagram.

(ii) Draw a network from the one-stage route matrix of
$$\begin{pmatrix} 4 & 2 & 1 \\ 2 & 0 & 3 \\ 1 & 3 & 2 \end{pmatrix}.$$

(iii) Find the two-stage route matrix of the diagram.

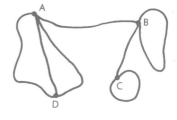

As we have seen and used to define a translation, a vector can be thought of as a list in a particular order. We used the top number to define horizontal movement, bottom number to represent vertical movement. (Written like this it is sometimes called a column vector.)

It follows then that the vector

$$\begin{pmatrix} 5 \\ 2 \end{pmatrix}$$

will represent a movement of 5 to the right and 2 up, no matter where you started. Hence, as seen on the diagram, the same vector can be in different places, but always parallel and of the same length.

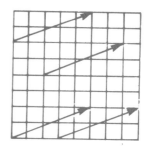

VECTOR ADDITION

We add and subtract vectors as the sum of the movements they represent. For example, if we *add* $\begin{pmatrix} 2 \\ 1 \end{pmatrix}$ to $\begin{pmatrix} 3 \\ -3 \end{pmatrix}$ then we can either do it as $\begin{pmatrix} 2 + 3 \\ 1 + -3 \end{pmatrix} = \begin{pmatrix} 5 \\ -2 \end{pmatrix}$ or graph both separately but following on, and seeing the single movement as in the illustration.

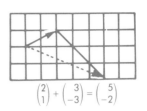

$$\begin{pmatrix} 2 \\ 1 \end{pmatrix} + \begin{pmatrix} 3 \\ -3 \end{pmatrix} = \begin{pmatrix} 5 \\ -2 \end{pmatrix}$$

VECTOR SUBTRACTION

If we wish to do a *subtraction*, say $\begin{pmatrix} 5 \\ 2 \end{pmatrix} - \begin{pmatrix} 3 \\ 4 \end{pmatrix}$, then we think about the $-\begin{pmatrix} 3 \\ 4 \end{pmatrix}$ as being moved back, hence this will become $+\begin{pmatrix} -3 \\ -4 \end{pmatrix}$ to give

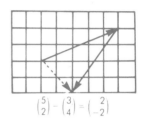

$$\begin{pmatrix} 5 \\ 2 \end{pmatrix} - \begin{pmatrix} 3 \\ 4 \end{pmatrix} = \begin{pmatrix} 5 \\ 2 \end{pmatrix} + \begin{pmatrix} -3 \\ -4 \end{pmatrix} = \begin{pmatrix} 2 \\ -2 \end{pmatrix}.$$

$$\begin{pmatrix} 5 \\ 2 \end{pmatrix} - \begin{pmatrix} 3 \\ 4 \end{pmatrix} = \begin{pmatrix} 2 \\ -2 \end{pmatrix}$$

or as seen graphically.

VECTOR NOTATION

Often these vectors are given a name, as in algebra, e.g. $\mathbf{a} = \begin{pmatrix} 2 \\ 1 \end{pmatrix}$. These letters are often written with a line or a squiggle under them, as a or a to denote they are vectors (in books they are usually printed in **bold** type), and if the vector is on a diagram with a line starting, say, with letter A and finishing with letter B, we would call the vector $\vec{AB}$. For example:

$$\vec{AB} = \begin{pmatrix} 3 \\ 2 \end{pmatrix} \quad \mathbf{a} = \begin{pmatrix} -2 \\ 1 \end{pmatrix} \quad \mathbf{b} = \begin{pmatrix} -1 \\ -3 \end{pmatrix}.$$

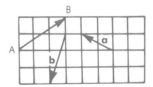

VECTOR MULTIPLICATION

We can *multiply* a vector by a number and this will denote another vector in the same direction but with a different length.

For example, if $\mathbf{a} = \begin{pmatrix} 2 \\ 1 \end{pmatrix}$ then $3\mathbf{a}$ will be

$$\mathbf{a} + \mathbf{a} + \mathbf{a} = \begin{pmatrix} 6 \\ 3 \end{pmatrix} \quad \text{or} \quad 3\mathbf{a} = 3 \times \begin{pmatrix} 2 \\ 1 \end{pmatrix} = \begin{pmatrix} 6 \\ 3 \end{pmatrix}.$$

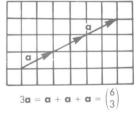

$$3\mathbf{a} = \mathbf{a} + \mathbf{a} + \mathbf{a} = \begin{pmatrix} 6 \\ 3 \end{pmatrix}$$

POSITION VECTORS

The *position vector* of a point is the column vector from the origin to that point.

For example, the position vector of point P with co-ordinates $(3, 6)$ will be $\begin{pmatrix} 3 \\ 6 \end{pmatrix}$ and probably written as **P**.

MAGNITUDE

The *magnitude* of a vector is generally shown by its length! Hence a column vector of $\begin{pmatrix} 4 \\ 3 \end{pmatrix}$ can be drawn as shown.

Then, by the rule of Pythagoras we can calculate the length of the vector to be 5 units.

EXERCISE 3

Where $\mathbf{a} = \begin{pmatrix} 5 \\ 4 \end{pmatrix}$, $\mathbf{b} = \begin{pmatrix} -1 \\ 3 \end{pmatrix}$, $\mathbf{c} = \begin{pmatrix} 2 \\ -1 \end{pmatrix}$, calculate the value of:

(i) $2\mathbf{a}$;

(ii) $3\mathbf{b} + \frac{1}{2}\mathbf{c}$;

(iii)   the magnitude of $\mathbf{a}$.

10 ▷ **MATRICES WITH TRANSFORMATIONS**

Most transformations, other than translations, can be described by a 2×2 matrix, and the simplest way to find this matrix is to consider what are known as *the base vectors*. These are the position vectors $\begin{pmatrix} 1 \\ 0 \end{pmatrix}$ and $\begin{pmatrix} 0 \\ 1 \end{pmatrix}$. When we find out where a transformation moves these base vectors to, then we can describe the transformation by their change.

WORKED EXAMPLE 5

Find the matrix which will describe a rotation of 90° clockwise around the origin.

Let us look at what happens to the base vectors $\begin{pmatrix} 1 \\ 0 \end{pmatrix}$ and $\begin{pmatrix} 0 \\ 1 \end{pmatrix}$ when rotating 90° clockwise.

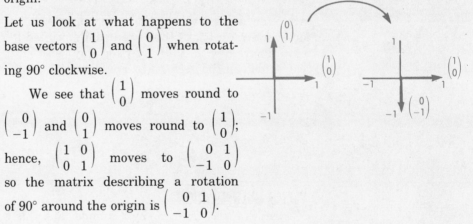

We see that $\begin{pmatrix} 1 \\ 0 \end{pmatrix}$ moves round to $\begin{pmatrix} 0 \\ -1 \end{pmatrix}$ and $\begin{pmatrix} 0 \\ 1 \end{pmatrix}$ moves round to $\begin{pmatrix} 1 \\ 0 \end{pmatrix}$; hence, $\begin{pmatrix} 1 & 0 \\ 0 & 1 \end{pmatrix}$ moves to $\begin{pmatrix} 0 & 1 \\ -1 & 0 \end{pmatrix}$ so the matrix describing a rotation of 90° around the origin is $\begin{pmatrix} 0 & 1 \\ -1 & 0 \end{pmatrix}$.

EXERCISE 4

Find the matrix that will describe a rotation of 180° around the origin.

MATRIX TRANSFORMATIONS

This gives quite a simple way of putting into matrix form a transformation. Now, for example, if we wish to program a computer to transform shapes on the screen we need a way of telling the computer the co-ordinates of a shape and calculating the new co-ordinates.

We do this by combining the transformation matrix with the position vectors of the given shape. For example, to see where the transformation matrix $\begin{pmatrix} 0 & 1 \\ -1 & 0 \end{pmatrix}$ will move (5, 3) we combine them in the following way, rewriting the co-ordinate as its position vector,

$$\begin{pmatrix} 0 & 1 \\ -1 & 0 \end{pmatrix}\begin{pmatrix} 5 \\ 3 \end{pmatrix} = \begin{pmatrix} 0 \times 5 + 1 \times 3 \\ -1 \times 5 + 0 \times 3 \end{pmatrix} = \begin{pmatrix} 3 \\ -5 \end{pmatrix}.$$

Try to be familiar with this notation, it is the usual way of asking the exam questions

Notice how the two combine by matrix multiplication, so the $(5, 3)$ will move to $(3, -5)$

The notation for a transformation T using this idea will often be written as:

$$T = \begin{pmatrix} x \\ y \end{pmatrix} \rightarrow \begin{pmatrix} a & b \\ c & d \end{pmatrix} \begin{pmatrix} x \\ y \end{pmatrix},$$

where the transformation matrix is $\begin{pmatrix} a & b \\ c & d \end{pmatrix}$ and you find the new position of any point (x, y) by substituting into the matrix multiplication.

WORKED EXAMPLE 6

A transformation R is defined by

$$R = \begin{pmatrix} x \\ y \end{pmatrix} \rightarrow \begin{pmatrix} -1 & 0 \\ 0 & -1 \end{pmatrix} \begin{pmatrix} x \\ y \end{pmatrix}.$$

Use this to transform the triangle shown in diagram (a) and hence or otherwise describe fully the transformation R.

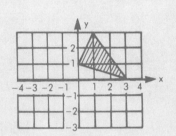

Combine the position vector of each vertex of the triangle to give:

$$\begin{pmatrix} 0 \\ 1 \end{pmatrix} \rightarrow \begin{pmatrix} -1 & 0 \\ 1 & -1 \end{pmatrix} \begin{pmatrix} 0 \\ 1 \end{pmatrix} = \begin{pmatrix} 0 \\ -1 \end{pmatrix};$$

$$\begin{pmatrix} 1 \\ 3 \end{pmatrix} \rightarrow \begin{pmatrix} -1 & 0 \\ 0 & -1 \end{pmatrix} \begin{pmatrix} 1 \\ 3 \end{pmatrix} = \begin{pmatrix} -1 \\ -3 \end{pmatrix}.$$

Check for yourself that $\begin{pmatrix} 3 \\ 0 \end{pmatrix} \rightarrow \begin{pmatrix} -3 \\ 0 \end{pmatrix}$.

This will give us the final position as shown in diagram (b). Do you recognise the transformation? It is a rotation of $180°$ around the origin.

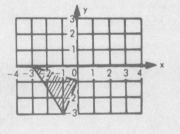

The above is also an enlargement of scale factor -1.

NOTE

You will generally find it easier to set down all the matrix multiplications in order. For instance, the above example could have been set out as

$$\begin{pmatrix} 0 & 1 & 3 \\ 1 & 3 & 0 \end{pmatrix} \rightarrow \begin{pmatrix} -1 & 0 \\ 0 & -1 \end{pmatrix} \begin{pmatrix} 0 & 1 & 3 \\ 1 & 3 & 0 \end{pmatrix} \begin{pmatrix} 0 & -1 & -3 \\ -1 & -3 & 0 \end{pmatrix},$$

done individually, but set out together.

COMBINATIONS OF TRANSFORMATIONS

If we *combine* two (or more) transformations that are both described by a matrix, then we can combine the matrices to give us the single matrix of the single combined transformation.

WORKED EXAMPLE 7

What is the matrix of the single transformation represented by reflecting in the y-axis, then enlarging with scale factor 2, centre of enlargement the origin.

The simplest way to find this matrix is, again, to consider the base vectors and find out what happens to them.

As you see, the transformation matrix will become $\begin{pmatrix} -2 & 0 \\ 0 & 2 \end{pmatrix}$.

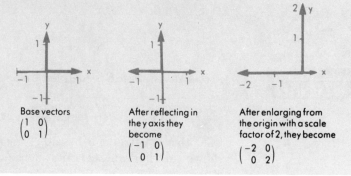

Base vectors
$\begin{pmatrix} 1 & 0 \\ 0 & 1 \end{pmatrix}$

After reflecting in the y axis they become
$\begin{pmatrix} -1 & 0 \\ 0 & 1 \end{pmatrix}$

After enlarging from the origin with a scale factor of 2, they become
$\begin{pmatrix} -2 & 0 \\ 0 & 2 \end{pmatrix}$

IDENTITY TRANSFORMATION

The *identity transformation* is the transformation that leaves a shape where it started; for example, a rotation of 360°, a reflection in the x-axis followed by a reflection in the x-axis, or even an enlargement of scale factor 1.

The matrix that describes the identity transformation is the one that describes the starting point of the base vectors, i.e. $\begin{pmatrix} 1 & 0 \\ 0 & 1 \end{pmatrix}$.

INVERSE TRANSFORMATION

An *inverse transformation* is a transformation that moves a shape back to its original position. For example, the inverse of 'reflection in the y-axis' will be 'reflection in the y-axis', the reflected shape will be reflected back to the original (self-inverse). Or, the inverse of 90° clockwise rotation around the origin will be a 90° anticlockwise (or 270° clockwise) rotation around the origin.

It follows also that if a matrix is used to describe a transformation then the inverse transformation is described with what we would call the inverse matrix. There is a complicated formula to find this inverse matrix which would always be given to you to use in an examination. The inverse matrix of $\begin{pmatrix} a & b \\ c & d \end{pmatrix}$ is given by

$$\frac{1}{(ad - bc)}\begin{pmatrix} d & -b \\ -c & a \end{pmatrix}.$$

(You see, I said it was complicated.) You will not need to learn this formula, it will always be given to you, but it is necessary to be familiar with it.

WORKED EXAMPLE 8

Find the matrix defining the inverse of the transformation F, where

$$F = \begin{pmatrix} x \\ y \end{pmatrix} \rightarrow \begin{pmatrix} 0 & 2 \\ -1 & 0 \end{pmatrix}\begin{pmatrix} x \\ y \end{pmatrix}.$$

We need the inverse of $\begin{pmatrix} 0 & 2 \\ -1 & 0 \end{pmatrix}$ and can use the formula quoted above, where $a = 0$, $b = 2$, $c = -1$, $d = 0$. Hence, the inverse is

$$\frac{1}{(0 - -2)}\begin{pmatrix} 0 & -2 \\ 1 & 0 \end{pmatrix} \quad \text{which is} \quad \frac{1}{2}\begin{pmatrix} 0 & -2 \\ 1 & 0 \end{pmatrix} \quad \text{which is} \quad \begin{pmatrix} 0 & -1 \\ \frac{1}{2} & 0 \end{pmatrix}.$$

Hence, the inverse transformation is defined as

$$\begin{pmatrix} x \\ y \end{pmatrix} \rightarrow \begin{pmatrix} 0 & -1 \\ \frac{1}{2} & 0 \end{pmatrix}\begin{pmatrix} x \\ y \end{pmatrix}.$$

EXERCISE 5

(i)   Find out the transformation matrix for a one-way stretch of scale factor 3 from the y-axis.

(ii)   Find the inverse transformation matrix in two different ways.

SOLUTIONS TO EXERCISES

S1

Your solution should be as below with no shapes overlapping.

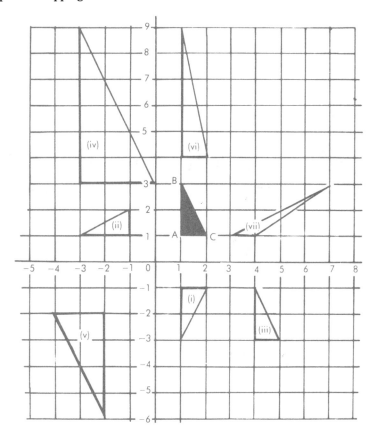

S2

(i)   The matrix will be:

$$\begin{pmatrix} 0 & 1 & 0 & 3 \\ 1 & 2 & 1 & 0 \\ 0 & 1 & 2 & 0 \\ 3 & 0 & 0 & 0 \end{pmatrix}$$

(ii)   A similar type of network as shown.

(iii)   Generally it is more accurate to find the two-stage route matrix by squaring the one-stage matrix. Here, the two-stage route matrix will be:

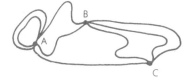

$$\begin{pmatrix} 10 & 2 & 1 & 0 \\ 2 & 6 & 4 & 3 \\ 1 & 4 & 5 & 0 \\ 0 & 3 & 0 & 9 \end{pmatrix}$$

S3

(i) $2\mathbf{a} = \begin{pmatrix} 2 \times 5 \\ 2 \times 4 \end{pmatrix} = \begin{pmatrix} 10 \\ 8 \end{pmatrix}$

(ii) $3\mathbf{b} + \tfrac{1}{2}\mathbf{c} = \begin{pmatrix} -3 \\ 9 \end{pmatrix} + \begin{pmatrix} 1 \\ -\tfrac{1}{2} \end{pmatrix} = \begin{pmatrix} -2 \\ 8\tfrac{1}{2} \end{pmatrix}$

(iii)   Magnitude $= \sqrt{(5^2 + 4^2)} = 6.4$.

S4

$\begin{pmatrix} 1 \\ 0 \end{pmatrix}$ will move to $\begin{pmatrix} -1 \\ 0 \end{pmatrix}$;

$\begin{pmatrix} 0 \\ 1 \end{pmatrix}$ will move to $\begin{pmatrix} 0 \\ -1 \end{pmatrix}$;

hence the matrix will be $\begin{pmatrix} -1 & 0 \\ 0 & -1 \end{pmatrix}$.

S5

(i)   By looking at base vectors, you will find that $\begin{pmatrix} 1 \\ 0 \end{pmatrix} \rightarrow \begin{pmatrix} 3 \\ 0 \end{pmatrix}$ and $\begin{pmatrix} 0 \\ 1 \end{pmatrix}$, which is on the invariant line, stays as $\begin{pmatrix} 0 \\ 1 \end{pmatrix}$;

hence, the matrix will be $\begin{pmatrix} 3 & 0 \\ 0 & 1 \end{pmatrix}$.

(ii)   Inverse transformation gives a one-way stretch of scale factor $\tfrac{1}{3}$ from the y-axis, hence $\begin{pmatrix} \tfrac{1}{3} & 0 \\ 0 & 1 \end{pmatrix}$.

OR, use the rule for finding inverse to give inverse $=$

$$\frac{1}{3 - 0}\begin{pmatrix} 1 & 0 \\ 0 & 3 \end{pmatrix} = \frac{1}{3}\begin{pmatrix} 1 & 0 \\ 0 & 3 \end{pmatrix} = \begin{pmatrix} \tfrac{1}{3} & 0 \\ 0 & 1 \end{pmatrix}.$$

EXAM TYPE QUESTIONS

Q1

(a) On the diagram reflect the letters AB in (i) the x-axis, (ii) the y-axis.

(b) Complete the diagram so that it has rotational symmetry of order 2.

(NEA)

Q2

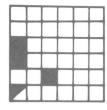

Blank crosswords often have two lines of symmetry, from each corner to corner diagonally. Complete the blank crossword given here in this way.

Q3

(a) On the diagram given, enlarge the rectangle
    (i) from ✱ with scale factor 3,
    (ii) from ⊙ with scale factor 2.

(b) What do you notice about the increase in area in each case?

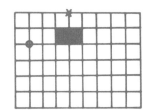

Q4

Show how the shape given will tessellate in a regular pattern. (Show at least 8 shapes.)

Q5

(a)   (i)   Draw on a grid, with both axes from -4 to 4, the triangle having vertices $A(1, 1)$, $B(1, 3)$ and $C(3, 0)$. Label this T.

     (ii)   Rotate T about $(0, 0)$ through $90°$ clockwise. Call this TT.

     (iii)   Reflect TT in the y-axis and label this TTT.

(b)   Describe fully the single transformation needed to transform TTT back to T.

Q6

(a) Describe fully the single transformations that will transform shape A to (i) B, (ii) C, (iii) D and (iv) E.

(b) Which shapes are congruent with A?

(c) Which shapes are similar to A (but not congruent)?

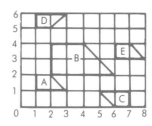

Q7

(a) Describe the network of roads between the towns shown as a route matrix.

(List them alphabetically in the matrix.)

(b) Draw a network that could be described with this given route matrix:

$$\begin{pmatrix} 4 & 0 & 1 \\ 0 & 0 & 2 \\ 1 & 2 & 6 \end{pmatrix}.$$

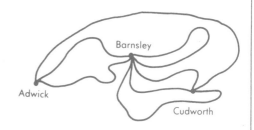

Q8

Where the vectors **a** and **b** are defined on the diagram, draw on the same diagram the vectors represented by (i) $3\mathbf{a} + 2\mathbf{b}$, (ii) $\mathbf{a} - \mathbf{b}$.

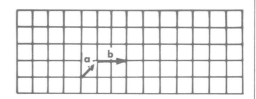

Q9

(a) On a grid with values from -5 to 5 on both axes, draw the triangle with vertices $A(1, 2)$, $B(4, 2)$ and $C(4, 1)$.

(b) Draw the image of ABC after the transformations determined by:

(i) $T: \begin{pmatrix} x \\ y \end{pmatrix} \rightarrow \begin{pmatrix} 1 & 0 \\ 0 & -1 \end{pmatrix}\begin{pmatrix} x \\ y \end{pmatrix}$;   (ii) $U: \begin{pmatrix} x \\ y \end{pmatrix} \rightarrow \begin{pmatrix} 0 & 1 \\ 1 & 0 \end{pmatrix}\begin{pmatrix} x \\ y \end{pmatrix}$

and label them t and u respectively.

(c) Describe as fully as possible the transformations T and U.

(d) Describe the single transformation that will move t to u.

Q10

(a) On a grid with values from 0 to 6 on both axes draw the rectangle with vertices $O(0, 0)$, $A(3, 0)$, $B(3, 2)$ and $C(0, 2)$.

(b) (i) Draw the image of the rectangle $OABC$ after the transformation where the transformation matrix is R when $R = \begin{pmatrix} 1 & 1 \\ 0 & 1 \end{pmatrix}$.

(ii) What has happened to the area of the transformed shape?

(c) Write down the matrix of the inverse transformation of R.

Q11

(a) On a grid with values from -4 to 4 on both axes, draw the triangle with vertices $A(0, 1)$, $B(1, 1)$ and $C(2, 0)$.

(b) Draw the image of the triangle ABC after the transformation defined by E, where $E: \begin{pmatrix} x \\ y \end{pmatrix} \rightarrow \begin{pmatrix} -2 & 0 \\ 0 & -2 \end{pmatrix}\begin{pmatrix} x \\ y \end{pmatrix}$.

(c) Describe fully this transformation.   (NEA)

Q12

Describe the transformation represented by the matrix $\begin{pmatrix} 4 & 0 \\ 0 & 1 \end{pmatrix}$.

Q13

The diagram illustrates a regular hexagon $ABCDEF$. Show by vectors that the length AD is twice the length BC.

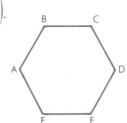

Q14

The transformation T consists of a reflection in the x-axis followed by an enlargement with centre $(0, 0)$ and scale factor 2. Find

(i)   the image of $\begin{pmatrix} 1 \\ 0 \end{pmatrix}$ and $\begin{pmatrix} 0 \\ 1 \end{pmatrix}$ under T;

(ii)   the matrix associated with T;

(iii)   the image of $\begin{pmatrix} 2 \\ 3 \end{pmatrix}$ under T.   (NEA)

Q15

A milkman delivers bottles of milk to four guest houses each day as follows:

Number 1: 27 gold top, 7 silver top.
Number 2: 13 silver top, 8 gold top.
Number 3: 16 silver top, 19 gold top.
Number 4: 23 silver top only.

(a) Copy and complete this matrix, M, to show these deliveries.

$$\begin{pmatrix} 27 & 8 & \ldots & \ldots \\ 7 & 13 & \ldots & \ldots \end{pmatrix}$$

(b) What is the order of matrix M?

(c) The matrix S is $(1, 1)$. Find R if $R = SM$.

(d) What information does R give?

(e) Gold top costs 22p and silver top costs 20p. Write this as a matrix C which can be multiplied by M.

(f) Find the product of C and M. What information does this give?

(g) Write a matrix, P, which can be multiplied by M to give the number of bottles delivered by the milkman for each type of milk, per day.

(h) Which combination of some or all of the matrices M, S, R, C, P will give the total money taken by the milkman? *YOU DO NOT HAVE TO CALCULATE THIS.*

OUTLINE ANSWERS TO EXAM QUESTIONS

A1

(a) You should end up with a diagram looking like this (but not with the dotted AB).

(b) Then fit the last AB into the bottom right-hand corner as we've illustrated with the dotted AB.

You would have found the question easier if you had used tracing paper.

A2

Draw in the diagonal lines of symmetry, then complete the crossword for one line of symmetry first then the other, to end up with the figure as shown. Here it is easiest to use the squares to help.

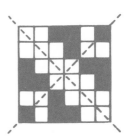

A3

(a) The enlargements should be as shown in this diagram. Use the lines on the grid to help you.

(b) You should notice that for an enlargement of scale factor 2 the area is 4 times as big (2×2), whilst for scale factor 3 the area is 9 times as big (3×3).

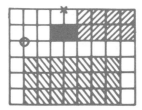

A4

The tessellation will need to fit together to form a regular pattern and leave no spaces, as in the diagram.

A5

(a) The diagram you should have drawn is shown here.

(b) The transformation back from TTT to T is a reflection in the line $y = -x$. If you gave an answer involving anything other than a reflection then you would get no marks.

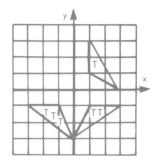

A6

(a) A to B is an enlargement of scale factor 2 with the centre of enlargement (0, 0).

A to C is a rotation of 180° around (4, 1).

A to D is a reflection in the line $y = 3\frac{1}{2}$.

A to E is a translation with a vector $\begin{pmatrix} 5 \\ 2 \end{pmatrix}$.

(b)   'Congruent' means all angles the same and corresponding lengths the same; hence, all except B will be congruent.

(c)   This leaves only B which is similar but not congruent.

A7

(a) You should have made a route matrix like this $\begin{pmatrix} 0 & 2 & 1 \\ 2 & 2 & 3 \\ 1 & 3 & 0 \end{pmatrix}$.

(b) The network could be very similar to that shown by the diagram, but the lines from, say, A to A could be like C to C and *not* inside each other.

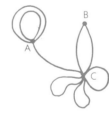

A8

(i) $3\mathbf{a} + 2\mathbf{b}$ is marked with a dotted line.

(ii) $\mathbf{a} - \mathbf{b}$ is also marked on the diagram, but any vector will do that is parallel to those marked.

A9

Shown opposite is the diagram you should have for (a) and (b).

(c) T is a reflection in the x-axis; U is a reflection in the line $y = x$.

(d) t moves to u with a 90° anticlockwise rotation around the origin.

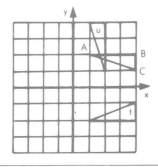

A10

Illustrated is the diagram you should have for (a) and (b)(i).

(b)(ii) The area remains the same, since the area of a parallelogram is base × height, which is 6, the same as the rectangle.

(c) Using the formula for an inverse matrix given in the text, when $a = 1$, $b = 1$, $c = 0$ and $d = 1$, we get:

the inverse matrix $= \dfrac{1}{(1 - 0)}\begin{pmatrix} 1 & -1 \\ 0 & 1 \end{pmatrix}$

which is $\begin{pmatrix} 1 & -1 \\ 0 & 1 \end{pmatrix}$.

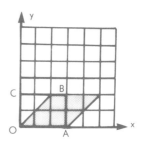

A11

Illustrated is the diagram you should have for (a) and (b).

(c) The transformation is an enlargement of scale factor -2, with the centre of enlargement being the origin.

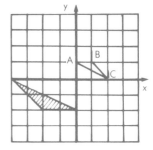

A12

Either by drawing your own shape and investigating what happens or by looking at base vectors, you should find that the transformation represents a one-way stretch of scale factor 4 with the y-axis invariant.

A13

Let $BC = \mathbf{b}$, then since it is a regular polygon, we can divide it up at the centre, O, as seen in the diagram. Now, AO is parallel and equal to BC, hence $\overrightarrow{AO}$ is equal to $\mathbf{b}$, as is $\overrightarrow{OD}$; hence $\overrightarrow{AD}$ is equal to $2\mathbf{b}$, twice the size of $\overrightarrow{BC}$. (There are other ways of doing this that would be just as valid and acceptable.)

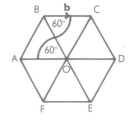

A14

(i) By possibly drawing on a grid these base vectors, we can see that

$\begin{pmatrix} 1 \\ 0 \end{pmatrix} \rightarrow \begin{pmatrix} 2 \\ 0 \end{pmatrix}$ and $\begin{pmatrix} 0 \\ 1 \end{pmatrix} \rightarrow \begin{pmatrix} 0 \\ -2 \end{pmatrix}$.

(ii) Hence the matrix will be $\begin{pmatrix} 2 & 0 \\ 0 & -2 \end{pmatrix}$.

(iii) $\begin{pmatrix} 2 & 0 \\ 0 & -2 \end{pmatrix}\begin{pmatrix} 2 \\ 3 \end{pmatrix} = \begin{pmatrix} 4 \\ -6 \end{pmatrix}$.

A15

(a) Completed from the information at the beginning of the question:

$$M = \begin{pmatrix} 27 & 8 & 19 & 0 \\ 7 & 13 & 16 & 23 \end{pmatrix}.$$

(If you got the last two columns upside down, look again at how the information is given, and how the matrix starts.)

(b) (2×4).

(c) $(1 \quad 1)\begin{pmatrix} 27 & 8 & 19 & 0 \\ 7 & 13 & 16 & 23 \end{pmatrix} = (34 \quad 21 \quad 35 \quad 23)$.

(d) The number of bottles that are delivered to each guest house each day.

(e) To be multiplied it must be a (1×2) since M has two rows. Hence, $C = (22 \ 20)$.

(f) $CM = (734 \ 436 \ 738 \ 460)$. This is the total cost each day to each guest house.

(g) This will have to be $P = \begin{pmatrix} 1 \\ 1 \\ 1 \\ 1 \end{pmatrix}$ and worked out as MP.

(h) $(CM)P$.

G R A D E   C H E C K L I S T

FOR A GRADE F

You should know:

what a transformation is, and what a reflection and an enlargement are;

what a tessellation is;

and be able to:

reflect a given plane shape in a given line;

enlarge any given plane shape with any positive scale factor;

find whether a shape will tessellate or not.

FOR A GRADE C

You should also know:

what a rotation and a translation are in a mathematical sense;

what a vector is;

and also be able to:

rotate a given shape around a point by 90° or 180°;

translate any shape with any given vector;

add and subtract vectors;

describe a network by a route matrix and draw a network from a given route
matrix.

FOR A GRADE A

You should also know:

when it is possible to add or subtract suitable matrices;

what a one-way stretch is;

what a transformation matrix and base vectors are;

what is meant by an identity and an inverse transformation;

you should understand:

what a transformation matrix actually does;

what is meant by an enlargement with a negative scale factor;

and be able to:

discover the transformation any transformation matrix represents;

describe simple transformations with a transformation matrix;

find the inverse of a transformation, and of a transformation matrix.

STUDENT'S ANSWER - EXAMINER'S COMMENTS

QUESTION

On graph paper provided draw the triangle ABC, where A, B, C are the points $(1,3), (2,3), (2,5)$ respectively.

The transformation P is a reflection in the line $x = -1$.

The tranformation Q is a reflection in the line $y = 1$.

R is the translation $\begin{pmatrix} 2 \\ -2 \end{pmatrix}$

(a) Triangle ABC is mapped onto triangle $A_1B_1C_1$ by P
     Triangle $A_1B_1C_1$ is mapped onto triangle $A_2B_2C_2$ by Q
     Triangle A_2B_2C is mapped onto triangle $A_3B_3C_3$

Draw these triangles on your diagram and label them clearly.

> **not a bad attempt. Yet because $A_3 B_3 C_3$ is wrong, there can be few marks gained here.**

(b) Describe the single transformation which maps triangle ABC onto triangle $A_3B_3C_3$.

Answer   *AN ENLARGED REFLECTION IN THE LINE X = −2½*

(c) A further transformation T is an enlargement, centre the origin, scale factor ½. Draw the image of triangle ABC under T.

(d) If the area of a triangle is x square units, state the area of the image of this triangle under T.

> **the correct answer of $\frac{1}{4}$ x is badly written here, and would score few marks**

Answer   *¼ x*

> **confused here. An enlargement has been drawn instead of a translation**

> **good**

> **good enlargement of scale factor ½**

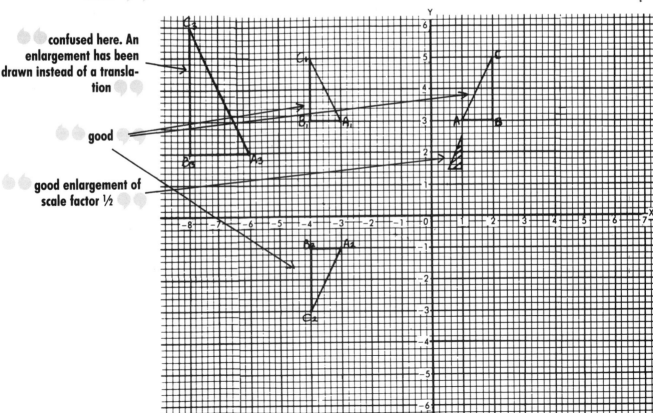

STATISTICS
AND
PROBABILITY

Statistics are all around us. They are used on television and in the papers, and sport is littered with them. However, the term *statistics* used here refers to more than pieces of data or information. It refers to the various *methods* of organising data, or of displaying data so that they make more sense.

You must be able to read the statistics as presented and to display that information in a way that highlights the major points or trends. The vast majority of examination questions will leave you in little doubt as to exactly what type of display you should use. You might also be asked for an interpretation or a conclusion to be drawn from the statistics.

The *probability* ideas you will meet are at a simple level and usually quite predictable. Do be careful of your notation. Many Examination Boards do not allow probability answers to be shown as anything but fractions (vulgar or decimal).

AVERAGE
FREQUENCY DISTRIBUTIONS
GROUPED DATA
CUMULATIVE FREQUENCY
SCATTER DIAGRAMS
PROBABILITY
TREE DIAGRAMS

USEFUL DEFINITIONS

Mode	the most frequently occurring value in a set of data
Median	the middle of a set of data in numeric order
Mean	a type of average calculated from all relevant data
Frequency	the number of times some defined event occurs
Pictogram	a display of information using pictures to represent the frequency
Bar chart	a display of information using bars of different lengths to represent the information
Pie chart	a circular picture divided in the ratio of the frequencies of the different events occurring
Histogram	like a bar chart, with the area of the bars being directly proportional to the frequency, often the widths of the bars will be different
Cumulative	increasing by successive additions
Ogive	the line representing cumulative frequency on a graph
Dice	(Often referred to as a die.) Unless told otherwise, any reference in a question to dice or die will mean a cube marked on each face from one to six spots (or the numbers). (Apologies to Dungeons and Dragons players!)

ESSENTIAL PRINCIPLES

What people usually mean by average is the 'middle thing', or the thing that most people do or have. But when we say that the average number of children in a family is 1.8, there is no family at all that has this number of children! So, let's look carefully at averages. There are three different types of average, and you should know the difference between them.

MODE

This is what 'most' people have or do. From a survey in which people were asked what their favourite evening drink was, the following information was gained:

Tea, 28;   Coffee, 115;   Cocoa, 136;   Other 9.

The most common choice was cocoa, hence we would say the *modal* drink was cocoa, or in this situation that the average evening drink was cocoa. (What is the average evening drink in your house?)

MEDIAN

This is the 'middle', once the information has been put into a specific order. For example, if you have seven people and wanted to find their median height you would put them into the order of their height and whoever was in the middle would be of median height (or average height).

WORKED EXAMPLE 1

Here are 15 test results, what is the median score?

(81, 63, 59, 71, 36, 99, 56, 31, 5, 65, 46, 83, 71, 53, 15)

Put the marks into order (5, 15, 31, 36, 46, 53, 56, 59, 63, 65, 71, 71, 81, 83, 99). Now find the middle one, which is 59. So the median score is 59.

If there is no single middle number then there will be two middle numbers, and the median in this case is halfway between the two middle numbers.

An easy way to work out where the middle is, is to put the numbers into order, count how many you've got (call this n), add 1 and divide by 2, i.e. $(n + 1)/2$. This will tell you how many to count along for your median.

WORKED EXAMPLE 2

Find the median of 1, 3, 4, 4, 6, 8, 8, 9, 10, 13.

Count the numbers, there are ten of them. Add one and divide by 2, giving $5\frac{1}{2}$, so count along five numbers and you come to the 6. Now you need halfway between this and the 8, which is 7, so the median of this list is 7.

MEAN

This is often known as the *arithmetic mean* and is perhaps the average that most people are familiar with and really intend by the word 'average'. It is found by adding up all the data and dividing by how many items of data you had. So, for example, the mean test score from Example 1 above is found by adding together all the scores and dividing by the total number of scores, which was 15. This will give us a mean of 834 ÷ 15, which is 55.6.

**WORKED
EXAMPLE 3**

G. Boycott scored the following number of runs in five test matches one year: 250, 85, 175, 110, 215. What was the mean number of runs scored in these matches?

Add up each score to give 835, divide this by 5 to give 167, which will be the mean number of runs scored per test match.

EXERCISE 1

From the numbers 2, 5, 1, 7, 1, 1, 4, state the (i) mode, (ii) median, (iii) mean.

RANGE

The *range* of a set of data is simply the difference between the highest and the lowest. For example, in G. Boycott's test scores in the above example, his range is 250 − 85, which is 165.

2 ▷ FREQUENCY DISTRIBUTIONS

Frequency is the number of times a certain event has happened, and is often found by means of a *tally chart*. For example, if you were doing a road survey and wished to count how many cars, buses, vans and lorries went past a particular spot during an interval of time it would be tedious and quite hard to count them all, so we use a tally chart.

The diagram shows an example of a tally made outside a school in a half-hour period one Friday morning. Each time a vehicle came by one person shouted out what it was, say, 'car', and someone else with the tally chart put a little mark (or tally) in the correct space besides the type of vehicle. Every fifth tally was put through the previous four (as you see). This was to make counting up easier and the figures were more likely to be accurate. Finally the frequency was found by adding up all the tally marks for each type of vehicle and putting this total in the end column marked *f* (for frequency). These are then added up to give the total number of vehicles passing. This way of collecting information (or data) is widely used in surveys for all sorts of things.

Once you have collected the frequencies, or been given them, you need to decide how to display this information. If the information is such that you can draw nice pictures of it, and you do not need to be all that accurate, then you can display the information on a pictogram, otherwise use a bar chart. However, in an examination situation you are most likely to be told which to do. In that case you simply need to decide upon a suitable scale for the frequency, since neither the pictogram nor the bar chart wants to be too large or too small.

	Tally	f
Cars	ЖЖ ЖЖ ЖЖ ЖЖ ЖЖ ЖЖ ЖЖ ЖЖ ЖЖ ЖЖ III	53
Buses	ЖЖ ЖЖ II	12
Vans	ЖЖ ЖЖ ЖЖ ЖЖ ЖЖ IIII	29
Lorries	ЖЖ ЖЖ ЖЖ I	16
	Total	110

PICTOGRAMS

The diagram shows a *pictogram*; it displays information with pictures. Here it is displaying information about a poll taken shortly before an election in a town. Note that a whole person represents 100 votes, and so we can use half a person (see the 'LOONY' party) to represent 50 votes. Here you see displayed 'LAB' with 300 votes and 'CONS' with 350.

Notice how a pictogram must have what is called a *key*, which tells us how many the individual 'pictures' stand for, and also that each 'picture' – here it is of a man – is of the same size.

This kind of display can be 'animated' – that is, 'come alive' – on the TV where changing information can be shown to walk about from one poll to the next.

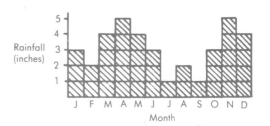

BAR CHARTS

This is a *bar chart*, each bar representing a piece of information so it can be more accurate than a pictogram since the scale can be made to have much smaller units. Often a bar chart will have gaps between the bars, but not necessarily. This bar chart displays how much rain fell each month (to the nearest inch) during one year at a holiday resort. The wettest months are April and November, both with 5 inches of rain. You should be able to pick out from this bar chart when the 'summer' was! Note how this bar chart has both axes fully labelled just like a normal graph. This is an important part of any bar chart as it helps us to interpret the displayed information.

PIE CHART

This is a *pie chart*, so called since it has the appearance of a pie and is cut into slices to illustrate the different 'ingredients' of the pie. This pie chart illustrates the gases in our air. The actual information is difficult to read accurately, but it does show us how the vast majority of the air is made up from nitrogen, only about a fifth being oxygen. The small shaded sector represents about 1% of what are called inert gases.

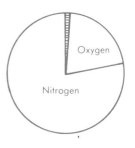

You are likely to be asked to show in an examination that you can extract some precise information from a pie chart.

**WORKED
EXAMPLE 4**

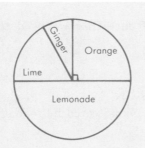

The pie chart represents the ingredients of 500 ml of a drink. There are 250 ml of lemonade and 25 ml of ginger.

(a)   How much (i) orange, (ii) lime, is there?

(b)   What angle represents ginger?

(a)   The orange is represented by an angle of 90°, that is $\frac{90}{360}$ of a circle, which is $\frac{1}{4}$; hence the orange is $\frac{1}{4}$ of 500 ml, which is 125 ml. The lime and ginger together form a right angle, and so are the same quantity as the orange, which is 125 ml. Since the ginger accounts for 25 ml, the lime will be 125 − 25, or 100 ml.

(b)   The angle for ginger will be a fraction of 360°. This fraction will be $\frac{25}{500}$, so on the calculator, calculate $\frac{25}{500} \times 360$, which is 18°.

CONSTRUCTION OF PIE CHARTS

To *construct a pie chart* there is a set way to go about it once you have gathered your information. For example, say we found the information in Table 1. Now we need to find the angle of the sector that each channel will be. We do this by finding what fraction of the whole data each channel is, and using this to find the same fraction of a complete circle or of 360°.

Table 2 illustrates what we have done and the completed pie chart. Note how, when the pie chart was being drawn, the very first angle to be put in would have been the smallest, then the next smallest and so on, so that the last angle should be the largest (hence any slight error is not so noticeable). The pie chart has also been fully labelled with the description of each sector and its angle.

TV viewing figures	
Channel	Number
BBC 1	3000
BBC 2	1000
ITV 3	500
ITV 4	7500

Table 1

Channel	Frequency	Angle
BBC 1	3000	$\frac{3000}{12000} \times 360 = 90°$
BBC 2	1000	$\frac{1000}{12000} \times 360 = 30°$
ITV 3	500	$\frac{500}{12000} \times 360 = 15°$
ITV 4	7500	$\frac{7500}{12000} \times 360 = 225°$
	12000	360°

Table 2

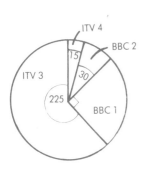

NOTE

Although the pie charts you see in everyday use will probably not have their sector angles labelled, it is usually expected in an examination situation where you are trying to show that you know what the angle should be.

EXERCISE 2

A school survey was done on 90 pet owners with the following numbers being the favourite pets:

Rabbit, 20;   Cat, 27;   Dog, 34;   Bird, 9.

Represent this information on (i) a bar chart and (ii) a pie chart.

HISTOGRAM

There are two types of data that we need to be able to distinguish between: *discrete* and *continuous*.

DISCRETE DATA

These are data that *can* be identified on their own by a single number. For example, the number of goals scored by teams, the number of people in a car or the marks given for a test.

CONTINUOUS DATA

These are data that *cannot* be measured exactly and are usually given to a rounded off amount. For example, the height of a number of people, or weight, a length of time or a person's age.

When we use continuous data we generally replace bar charts with histograms, which definitely do not have any gaps in them because of the continuous nature of the data included.

This histogram illustrates the age incidence of duodenal ulcers found during a survey in a city hospital. Note a difference from the bar chart in the way the horizontal axis is labelled. Since each item of data is continuous, each 'bar' has not been labelled as such, although its end points have.

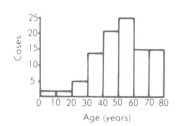

In a histogram it is strictly the 'area' of the bars (rectangles) that should distinguish the frequencies. However, at all but the highest level examination, your bars will all be of equal width and hence the heights will still be used to determine the frequencies, as in the example given.

---

3 ▷ GROUPED DATA

Quite often we are faced with non-precise information, as in the following example. A survey was carried out to find the average height of pupils in a school, and this table represents the results.

Height (cm)	Frequency
100–109	12
110–119	19
120–129	35
130–139	32
140–149	21
150–159	10
160–169	2

Table 14.3

The data have been *grouped* to make the information more easily read, and the raw data will have been rounded off. For example, if someone had a height of 109.8 cm, then it seems that he or she does not appear on the chart, but 109.8 will have been rounded off to give 110 cm and put into that particular group.

It is an essential feature of a grouped frequency such as this that we know what the group 'boundaries' are. Here, for example, the first group will really take in heights from 99.5 to 109.499. . . (note that 109.5 would be rounded to 110). But in the table it is more convenient to show the heights in the way we have done. Care must be taken that we do not have any confusion at group boundaries. For example, if we had groups here of 100–110, 110–120, 120–130, etc., then it is not at all clear what happens at 110, 120, 130, etc., so this type of grouping must be avoided (although you'll see it in the Press and on TV).

USES OF GROUPED DATA: ESTIMATION OF THE MEAN

This kind of information can be put to good use by giving us nice compartments in which to draw up a histogram or pie chart if we so wish. One main use of this kind of information is to estimate what the 'mean' is. To do this we find the halfway mark of each group and estimate all frequencies in that group to have the same mark. For example, using the information in the previous item, to estimate the mean we would extend the table as follows:

Height (cm)	Halfway (m)	f	m × f
100–109	104·5	12	1254
110–119	114·5	19	2175·5
120–129	124·5	35	4357·5
130–139	134·5	32	4304
140–149	144·5	21	3034·5
150–159	154·5	10	1545
160–169	164·5	2	329
Totals		131	16999·5

Table 14.4

> Students often make silly errors in calculating these mid-values. Take great care.

The halfway point was found by calculating the mean of the group boundaries. Then the final column is the estimated total height of people in that group, found by multiplying halfway by frequency. Then we divide the total estimated height by the total frequency to give $16\,999.5 \div 131$, which will round off to 130 cm. It must be remembered that this is an *estimated mean* height and therefore has been rounded to three significant figures.

EXERCISE 3

The grouped frequency table below shows the number of candidates obtaining various ranges of marks in an examination. Complete this table and hence estimate the mean mark per candidate (give your answer to the nearest integer).

Mark	Mid-valve	Frequency	
0–20		60	
21–50		680	
51–80		1160	
81–100		200	

Table 14.5

UNEQUAL WIDTH HISTOGRAMS

Do you remember that it is the area of the rectangles in a histogram that represent the data? This is used when we consider *unequal interval histograms* with, now, the vertical axis being *frequency density* instead (often abbreviated to f.d.), so that the reading of the frequency density multiplied by the width of the rectangle will give the actual frequency which that particular rectangle is representing.

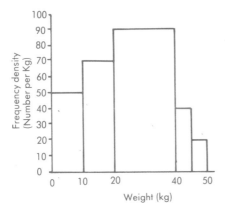

For example, in this histogram the weight of an 'average' sized box handled by a book delivery service is between 20 and 40 kg. The histogram illustrates the number of boxes handled in one particular week.

Notice how, since the vast majority of the boxes are of average weight, they have been 'lumped' together and the smaller number of heavier weights have been made such that we can see the difference between just above average and beyond. So from this histogram the groupings are as follows:

Box weights	f.d. × width	Frequency
0–10	50 × 10	= 500
11–20	70 × 10	= 700
21–40	90 × 20	= 1800
41–45	40 × 5	= 200
46–50	20 × 5	= 100

So, as you can see, this is quite a good way to display information which does vary quite a bit.

EXERCISE 4

The weekly take-home wages of a firm's employees were given in a table as below, to show the spread of wages.

Wage per week	Number of employees
Under £50	8
Between £51 and £200	36
Between £201 and £225	4
Between £226 and £250	1

Draw a suitable histogram to illustrate this distribution.

4 ▷ CUMULATIVE FREQUENCY

Sometimes known as running total graphs. They can be quite useful to show a spread of marks and to find the median as well as quartiles. For example, the examination results for mathematics in one particular college one year were summarised as shown in the table.

Marks	Frequency	Cumulative frequency
0–20	50	50
21–30	50	100
31–40	95	195
41–50	120	315
51–60	90	395
61–70	55	450
71–80	30	480
81–100	20	500

Table 14.6

Notice how the *cumulative frequency* is a running total of the frequency given. On its own, this information is not particularly useful but it can be when we graph it on what we would call a cumulative frequency curve. This example graphs in the following way. The cumulative frequency is always put on the vertical scale, and we plot the groups near the boundary with the cumulative frequency, so that we can draw in the curve as shown, known as the *ogive*. It has a distinctive shape as here, and from this type of curve we can find quartiles and the interquartile range.

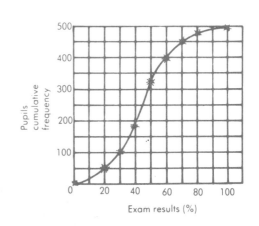

QUARTILES

As the name suggests, *quartiles* are found by dividing the cumulative frequency (c.f.) into four quarters. The points on a c.f. that give us the quartiles can be found by dividing the total into four equal groups. For example, imagine the graph that we used in the previous item split into 4. Since we have 500 pupil, the halfway mark is 250 (O.K. it actually is $250\frac{1}{2}$, but can you really tell the difference?)

This will give us the median mark (in our case about 44%). If we then halve the frequency again, or find one-quarter of the total frequency, it will give us 125. This is the quarter mark, or, as it is called, the *lower quartile*. In the same way find three-quarters of the way up the c.f. – here it will be 375 – and you find the *upper quartile* (which here is 56%).

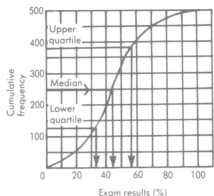

Careful, though; although this method works well for large frequencies, if you have smaller frequencies then you will need to be more precise. To find the exact median from a c.f. of n, you look for the mark of the $\frac{1}{2}(n + 1)$ on the c.f. To find the lower quartile you look for $\frac{1}{4}(n + 1)$ on the c.f., and so the upper quartile will be the $\frac{3}{4}(n + 1)$ on the c.f. But this only needs doing if it's going to make a significant difference on your ogive.

THE INTERQUARTILE RANGE

This is the difference between the upper and lower quartiles, and should be expressed simply as the number difference on the horizontal axis. For instance, in our example above, the interquartile range will be 56% – 33%, which is 23%. This tells us the spread of the middle half of the given population. This spread of 23% is not a very good one at all since it means that only 23% separates half the pupils at this college!

SEMI-INTERQUARTILE RANGE

This is exactly what it says: half of the interquartile range.

EXERCISE 5

A café owner kept records of the number of meals served each week over a period of time and summarised the figures in the following cumulative frequency table:

(a) Draw the ogive from this information.

(b) Use your ogive to estimate (i) the median number of meals served per week, and (ii) the interquartile range.

(c) The café runs at a loss if less than 180 meals are served in a week. For how many weeks did the café run at a loss?

Number of meals served per week	Number of weeks
Less than 100	20
Less than 200	90
Less than 300	240
Less than 400	370
Less than 500	435
Less than 600	470
Less than 700	490
Less than 800	500

Table 14.7

SCATTER DIAGRAMS

A *scatter diagram* is used to test for any relationships that may be present in your statistics. For example, we have 'scattered' the information about some children's heights and weights.

> You can draw a line on the diagram that represents the line of best fit. This will help you to see what the connection is.

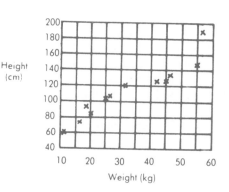

Name	cm	kg
James	135	47
John	105	26
Joseph	60	11
Paul	101	25
Michael	85	20
Jenny	74	16
Robert	120	32
Helen	130	45
Neil	130	42
Kirsty	95	18
Gary	180	57
Mark	145	55

Table 14.8

Each cross represents an individual child by plotting a co-ordinate found from (weight, height). From the chart we can easily deduce that the taller a child is the more likely he/she is to weigh more than a smaller child! You can use this scatter diagram to suggest very roughly the expected weight of a child of height 140 cm. You will see from the diagram that that child is likely to weigh more than 35 kg and less than 55 kg – not a very good estimation, but it can be used as a guide.

The use of scatter diagrams is one way of examining data to check if there is any type of connection between two properties of the same person (or whatever) in which case the points would lie in some line, as above. It is a practice in common use within examinations to see whether any one examination can be described as much too hard or easy compared with other subjects in which the same people are being examined.

6 ▷ PROBABILITY

> It's no use understanding this if you can't do fractions! I hope you can.

Or chance. This is finding out whether an event is likely or unlikely to happen, or whether one event has more chance of occurring than another.

We calculate the probability of an event happening as a fraction given by:

$$\frac{\text{the number of ways the event can happen}}{\text{the total number of ways that equally likely events can happen}}.$$

For example, the probability of rolling a dice and getting a number 5 is $\frac{1}{6}$ since there is only one 5 on the dice and six numbers altogether. Another example is the probability of cutting a pack of cards and obtaining an ace. The probability of this is 4/52 since there are four aces and 52 cards to choose from.

There are three important chances to know. An event that we may say has an 'even' chance or a '50–50' chance is one like tossing a coin and getting a head, the probability is $\frac{1}{2}$. The probability of something impossible is 0. For example, the probability of rolling a dice and getting a 7 is 0/6 which is 0. The probability of a certainty is 1. For example, the probability of rolling a dice and getting a number less than 7 is 6/6 which is 1.

Hence, any probability will be in between 0 and 1; the closer to nought then the more unlikely the event is to happen; the closer to 1 then the more likely the event is to happen.

WORKED EXAMPLE 5

In two bags of sweets, the larger one had 14 toffees and 6 mints, and the smaller one had 11 toffees and 4 mints. Being given a sweet out of a bag at random, which bag is more likely to give you a toffee than a mint?

The probabilities of getting toffees are: from the big bag $\frac{14}{20}$ and from the small bag $\frac{11}{15}$ We need to know which is the bigger fraction, and the easiest way to do this is to convert each fraction to a decimal number. $\frac{14}{20}$ is $14 \div 20$ which is 0.7, and $\frac{11}{15}$ is $11 \div 15$ which is 0.7333..., so $\frac{11}{15}$ is bigger than $\frac{14}{20}$. Therefore the small bag gives the best chance of getting a toffee.

EXPECTATION

If, for example, we know that the probability of rolling a 3 on a die is $\frac{1}{6}$, then if the die was rolled 100 times, how many times would we *expect* it to be a 3? The answer is found by multiplying the number of times you do it by the probability, so here it is $100 \times \frac{1}{6} = 16.66...$, which will be rounded to 17. Hence, you would expect to get around 17 threes if you rolled the die 100 times. (Try it, you'll not be far out.)

WORKED EXAMPLE 6

A firework being dud has a probability of $\frac{3}{100}$. Out of a big box of 450, how many duds would you expect?

Just multiply 450 by $\frac{3}{100}$ to get 13.5, which you would round to 14. So 14 dud fireworks would be expected.

EXERCISE 6

A doctor did a check on 140 patients chosen at random, and found that 42 of them had back trouble.
(i) State as a fraction in its simplest form the probability of any one of his patients having back trouble.
(ii) The doctor had 900 patients. How many of them would he expect to have back trouble?

COMBINED EVENTS

When we want to find the probability of a *combined event* – that is, where two or more events are happening at the same time – then we need to be clear about whether we want two events to happen at the same time or whether either event can happen but not necessarily at the same time as the other. These two fall into different types that can be described as 'AND' and 'OR'.

AND

AND is the type where both events happen at the same time. To find this combined probability we *multiply* the probabilities of each single event.

WORKED EXAMPLE 7

When rolling a die and tossing a coin, what is the probability that you will roll a 4 and toss a head at the same time?

The probability of rolling a four is $\frac{1}{6}$; the probability of tossing a head is $\frac{1}{2}$; hence, the combined probability of a four AND a head will be $\frac{1}{6} \times \frac{1}{2} = \frac{1}{12}$.

WORKED EXAMPLE 8

Find the probability of tossing a coin and getting a head four times in a row.

We are looking for tossing a head AND then a head AND then a head, and so on for four times. Hence, the probability will be $\frac{1}{2} \times \frac{1}{2} \times \frac{1}{2} \times \frac{1}{2}$, which is $\frac{1}{16}$.

OR

OR is the type when either one event or the other can happen but *not both at the same time*. To find this combined probability we *add* the probability of each event.

WORKED EXAMPLE 9

To finish a game a boy needed to throw a 6 or a 4. What was his chance of finishing on the next throw?

Each event has a probability of $\frac{1}{6}$, so for either event the probability will be $\frac{1}{6} + \frac{1}{6}$ which is $\frac{2}{6}$ or $\frac{1}{3}$.

WORKED EXAMPLE 10

On any day in February the weather probabilities are given as rain $\frac{1}{10}$, snow $\frac{1}{3}$. What is the probability of a day in February being either rainy or snowy? (We assume it cannot be both.)

Since we want the probability of either rain OR snow, then we add the probabilities of $\frac{1}{10}$ and $\frac{1}{3}$ to get $\frac{13}{30}$.

Tree diagrams are a very useful way of looking at all the possibilities of a given situation and help us to find various probabilities. It uses both ideas of AND and OR we have just met.

For example, when travelling down from Sheffield to Cornwall on the motorways there is a $\frac{1}{10}$ chance of being held up on the M1, and a $\frac{1}{5}$ chance of being held up on the M5. What are the chances of being held up, or not, on the total journey?

We can illustrate the possibilities with a tree diagram, as shown.

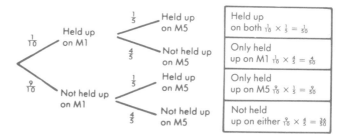

Notice how the individual probabilities have been put on the 'branches' of the tree diagram, with each 'pair' adding up to 1, since in each pair one of the branches *must* happen. Then to find the probabilities of the combined events, you just multiply along the branches as shown.

But, beware of the following type of situation.

WORKED EXAMPLE 11

I have a bag of sweets containing 5 toffees, 3 mints and 2 jellies. What are the probabilities of taking two of the same sweets out to eat?

You must consider taking out first one sweet, then the other, but when you take out the second sweet there are fewer sweets in the bag, so look carefully at the probabilities this situation gives us.

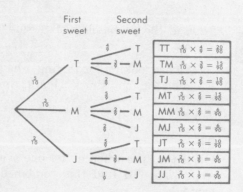

Notice again that each set of branches from the same point will all add up to 1. Look at the second sweet branches. There are only 9 sweets left in the bag and when the toffee was the first sweet there were only 4 toffees left, or in the case where the mint was the first sweet out there were 5 toffees and 2 jellies but only 2 mints. Also if you add up the final probabilities they should add up to 1.

WORKED EXAMPLE 12

Caroline has 30 cassettes, of which 15 are 'heavy metal', 10 are 'rock' and 5 are 'pop'. She is late for a party and asks her Mum: 'Oh, get me any two, Mum!!' Draw a tree diagram to help you find the probability that both cassettes are of the same type.

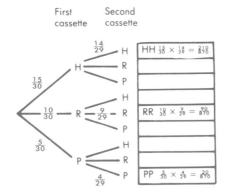

We shall not need all the probabilities for this particular question, so we shall only put the relevant probabilities on the tree diagram (but just test yourself that you could do it all if asked). Notice that by not cancelling down it keeps all the fractions the same type which helps when it comes to adding them together. So the probability of either both being heavy metal OR both being rock, OR both being pop will be:

$$\frac{210}{870} + \frac{90}{870} + \frac{20}{870} \quad \text{which is} \quad \frac{320}{870} \quad \text{(not far off a } \tfrac{1}{3} \text{ chance).}$$

SOLUTIONS TO EXERCISES

S1

(i)   Mode = 1;   (ii) median = 2;
(iii) mean = 21 ÷ 7 = 3.

S2

(ii) The angles in your pie chart should be:

Rabbit $\dfrac{20}{90} \times 360 = 80°$;

Cat = 108°;
Dog = 136°;
Bird = 36°.

S3

Care was needed with the mid-values, but then the estimated mean is

118 820 ÷ 2100 = 56.58.

To the nearest integer then, this is 57.

Mark	Mid-value (m)	Frequency (f)	$m \times f$
0–20	10	60	600
21–50	35·5	680	24 140
51–80	65·5	1160	75 980
81–100	90·5	200	18 100
Totals		2100	118 820

S4

Be careful with units and frequency density units, but you should have an answer like the one shown.

Notice how 1 'square' represents 1 person, hence the frequency density is 1 unit per £25, since the width of a square representing 1 person is £25. An alternative scale to the graph would be to have said 1 square represents 1 person.

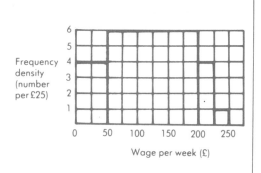

Frequency density (number per £25)

Wage per week (£)

S5

(a) When you graph this it should look something like:

(b) (i) Read from your graph, but the median is at 250 on the cumulative frequency, which is about 313 meals served.

(ii) The interquartile range is the difference of the upper and lower quartiles which should be around 407 − 223, which is about 184. (*Note*: in the examination you will be marked from *your* graph.)

(c) You need to read off from where meals served are 180, and read along the ogive to the c.f. This will give around 70 weeks that the café ran at a loss.

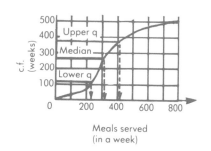

c.f. (weeks)

Meals served (in a week)

S6

(i) $\frac{42}{140}$ cancels down to $\frac{3}{10}$ or 0.3.

(ii) $900 \times \frac{3}{10} = 270$.

EXAM TYPE QUESTIONS

Q1

This bar chart shows how many books were sold at a shop each day during one week.

(a) How many books were sold altogether in that week?

(b) Which day was most likely to be half-day closing?

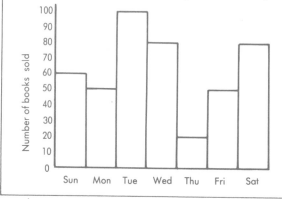

Q2

In a game of snakes and ladders I was on square 24, as shown on the diagram.
What would be the chance of my next throw of the die landing me at:
(a)   the bottom of a ladder?
(b)   the head of a snake?

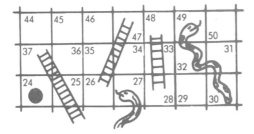

Q3

The table gives information about the weather one Friday in August for a selected number of towns in England.

(a)   What is the mean number of hours of sunshine?

(b)   What is the median maximum temperature (°C)?

(c)   What would be considered the 'average' weather?

Reports for the 24 hours ended 6 pm yesterday:

	Sun-shine hrs	Rain in	Max temp C	F	Weather
ENGLAND					
Birmingham	2.8	—	18	64	Sunny
Bristol	10.9	—	19	66	Sunny
Carlisle	1.4	—	17	63	Bright
London	6.7	—	22	72	Sunny
Manchester	0.6	.01	16	61	Showers
Newcastle	1.5	.10	16	61	Showers
Norwich	3.5	—	19	66	Bright
Nottingham	2.5	—	17	63	Sunny
Plymouth	8.9	—	16	61	Sunny

Q4

A survey in a hospital was carried out to see how long it took to give patients a check-up. The results are summarised in the table.

(a) By suitably extending the table, calculate the estimated mean time taken to give one patient a check-up. Give your answer to the nearest minute.

(b) A doctor is available to do check-ups on one afternoon for 4 hours. Using the average time per patient that you found in (a), estimate how many patients the doctor can see during this time.

(c) Using the average time per patient as before, estimate how many doctors the hospital would need in order to give 83 patients a check-up during a 3 hour morning session.

Time taken to the nearest minute	Frequency
11–15	18
16–20	27
21–25	12
26–30	3
Total	60

Q5

The probability of a train arriving early at a station is $\frac{1}{10}$. The probability of a train arriving late at a station is $\frac{2}{5}$.

(a)   If 400 trains are expected at a station during the day, how many of them are likely to arrive at the correct time?

(b)   What is the probability that both the trains arriving at the station from Exeter are late?
(SEG)

Q6

Pop stars often say they do not get a fair share of the selling price of the LP records they make. For example, for a record selling at £5.40: £1.50 goes to the shop, 90p goes to the government in tax, 54p is the cost of materials, £1.80 goes to the record company, and the rest goes to the pop star.

(a)   How much of the £5.40 goes to the pop star?

(b)   How many records must be sold for the pop star to be paid £1000?

(c)   Draw a suitable pie chart to show how the money paid for a LP record is distributed.

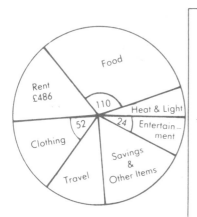

Q7

(a) In 1975, an apprentice electrician's 'take home' pay was £30 per week. His weekly budget was as follows:

Rent, food, heat and light   £9
Clothes   £6
Entertainment   £8
Travel   £4
Savings and other items   £3

Draw a pie chart to represent his weekly budget.

(b) The pie chart represents the 'average family' budget in 1975. The 'average family's' net income in 1975 was £3240. Calculate:

    (i)   how much was spent on food,
    (ii)   what angle is represented by rent,
    (iii) what percentage of the family's net income was spent on entertainment.

(c) By comparing the two pie charts, comment briefly on the major differences between the 'average family' budget and the apprentice's budget.

(SEG)

Q8

In a game to select a winner from three friends Arshad, Belinda and Connie, Arshad and Belinda both roll a normal die. If Arshad scores a number greater than 2 *and* Belinda throws an odd number, then Arshad is the winner. Otherwise Arshad is eliminated and Connie then rolls the die. If the die shows an odd number Connie is the winner, otherwise Belinda is the winner.

Calculate the probability (P) that
(i)   Arshad will be the winner,
(ii)   Connie will roll the die,
(iii) Connie will be the winner.   (LEAG)

Q9

The table below shows the prices of some paperback novels and the number of pages in them.

PRICE	85p	£1.00	95p	£1.25	£1.50	£1.65	95p	£1.00	£1.35	65p	75p
PAGES	224	254	170	236	330	380	210	190	320	136	150

On graph paper construct a scatter diagram for this information. Use scales of 2 cm to represent 50 pages and 2 cm to represent 20p.
(i)   Draw a line of best fit.
(ii)   Use your line to estimate the cost of a book with 300 pages.   (NEA)

Q10

To be an advanced motor mower driver you must pass three tests, and the following are the probabilities of passing these individual tests:

    Ability to walk in a straight line $\frac{3}{5}$
    Ability to swerve out of the way of stray animals $\frac{1}{5}$
    Ability to talk constantly while driving $\frac{9}{10}$

Each test is taken in the order given above.

(a) Draw a tree diagram to illustrate the ways in which you can fail and hence find the probabilities that
    (i)   you will fail only one test,
    (ii)   you fail to become an advanced motor mower driver.
(b) If 100 people were tested, how many would you expect to pass.

OUTLINE ANSWERS TO EXAM QUESTIONS

A1

(a)   Add the totals for each day and you should get 440.

(b)   The smallest number of books sold was on Thursday – a good answer for the half day.

A2

(a)   There are 2 ladders I can land at the bottom of out of a possible 6 squares, hence the probability is $\frac{2}{6}$ or $\frac{1}{3}$.

(b)   There is only 1 snake's head out of 6, so the probability is $\frac{1}{6}$.

A3

(a)   Add up all the hours and divide by 9 (since there are 9 towns in the list), and you will get 38.8 ÷ 9, which rounds off to a mean of 4.3 hours.

(b)   Put the nine temperatures in order and you will get (16, 16, 16, 17, 17, 18, 19, 19, 22) of which the middle one is 17, hence the median temperature is 17 °C.

(c)   The only average we can look for in the weather column is the mode, which is the one that appears most times. This is 'sunny'.

A4

You need a halfway mark then a $m \times f$ column, so the table should be extended as shown.

Fig. 14S.4

Time taken	Halfway (m)	f	$m \times f$
11–15	13	18	234
16–20	18	27	486
21–25	23	12	276
26 30	28	3	84
Total		60	1080

(a)   Hence the estimated mean will be 1080 ÷ 60, which is 18 minutes.

(b)   4 hours is 4 × 60 minutes, which is 240. So the doctor sees 240 ÷ 18, which is 13.33; hence he will only have time for 13 check-ups.

(c)   83 patients at 18 minutes each means 1494 'patient minutes' to be checked in 3 × 60 = 180 minutes. The hospital will therefore need 1494 ÷ 180 doctors, which is 8.3; so to see all the patients this needs to be rounded off to 9 doctors.

A5

(a)   The probability of the train arriving early OR late is $\frac{1}{10} + \frac{2}{5} = \frac{5}{10}$; hence, the probability that it is on time is $1 - \frac{5}{10} = \frac{5}{10}$; hence, you would expect $400 \times \frac{5}{10} = 200$ to arrive on time.

(b) $\frac{2}{5} \times \frac{2}{5} = \frac{4}{25}$.

A6

(a)   Add up the totals given and subtract from £5.40, and you will get £5.40 − £4.74, which is £0.66 or 66p.

(b)   To make £1000 you will need to sell £1000 ÷ £0.66 records which is 1515.15. . ., hence it would be necessary to sell 1516 records to be sure of £1000.

(c)   The angles for the pie chart will be found by the following: change everything to pence to simplify matters, then find each as a fraction of 360°.

Money		Angle	
Shop......................	150	$\dfrac{150}{540} \times 360$	$= 100°$
Tax	90	$\dfrac{90}{540} \times 360$	$= 60°$
Materials..............	54	$\dfrac{54}{540} \times 360$	$= 36°$
Company..............	180	$\dfrac{180}{540} \times 360$	$= 120°$
Pop star...............	66	$\dfrac{66}{540} \times 360$	$= 44°$
Total.....................	540		$360°$

Then the accuracy of the actual pie chart is up to you, but you should put the angles in starting with the smallest angle of 36° first and going up in order to 120°. Do not forget to label angles and sectors.

A7

(a) Rent, etc.: .... $\dfrac{9}{30} \times 360 = 108°$

   Clothes: .... $\dfrac{6}{30} \times 360 = 72°$

   Entertainment: .... $\dfrac{8}{30} \times 360 = 96°$

   Travel: .... $\dfrac{4}{30} \times 360 = 48°$

   Savings, etc.: .... $\dfrac{3}{30} \times 360 = 36°$

(b) (i) $\dfrac{110}{360} \times 3240 = £990$;

   (ii) $\dfrac{x}{360} \times 3240 = 486$;

     $x = \dfrac{486 \times 360}{3240} = 54°$

   (iii) $\dfrac{24}{360} \times 100 = 6.67\%$.

(c) Apprentices spend more on entertainment and clothes and much less on rent, food, heat and light.

A8

(i) $P(\text{A no.} > 2) = \dfrac{4}{6}$ or equiv.

   $P(\text{B odd no.}) = \dfrac{1}{2}$

   $P(\text{both above}) = \dfrac{2}{3} \times \dfrac{1}{2}$;

   $P(\text{A winner}) = \dfrac{1}{3}$.

(ii) $P(\text{C rolls die}) = 1 - (\text{i}) = \dfrac{2}{3}$.

(iii) $P(\text{C odd no.}) = \dfrac{1}{2}$;

   $P(\text{C winner}) = \dfrac{2}{3} \times \dfrac{1}{2} = \dfrac{1}{3}$.

A9

The answer to (ii) is taken from *your* line of best fit. But the answer should be between £1.25 and £1.40.

A10

You've been asked to draw a tree diagram, so you might lose marks if you tried the question without one! The tree diagram should look like the one here. Only the final probabilities that would be needed for the answers have been inserted (see if you can fill in the rest for yourself).

(a) (i) Only failing one test can be done in the 3 ways: PPF, PFP or FPP, add together their probabilities to get $\frac{3}{250} + \frac{108}{250} + \frac{18}{250}$, which is $\frac{129}{250}$.

    (ii) You fail to become a proper driver only if you do not pass all three tests, hence the probability is $1 - \frac{27}{250}$, which is $\frac{223}{250}$.

(b) If 100 people take the tests, then since the probability of passing all three is $\frac{27}{250}$ we would expect to see $100 \times \frac{27}{250}$ pass, which rounds to 11 people.

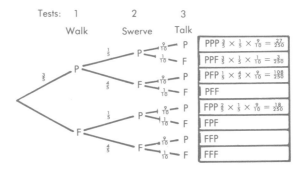

G R A D E   C H E C K L I S T

FOR A GRADE F

You should understand:

    the differences between the different types of average;

and be able to:

    display data by means of a bar chart or pictogram;

    read information from a bar chart, pictogram or pie chart, giving reasons for features observed;

    identify the mode, the median and the mean from a set of data;

    express probability as a simple fraction.

FOR A GRADE C

You should also know:

    when to use a histogram instead of a bar chart;

    what a scatter diagram is;

and also be able to :

    construct pie charts and histograms from data, including grouped data;

    estimate a mean from a grouped frequency table;

    use a tree diagram to find a set of probabilities;

    use a scatter diagram to check for any relationships between sets of data.

FOR A GRADE A

You should also understand:

    what a cumulative frequency and interquartiles are;

and also be able to:

    draw cumulative frequency curves and extract information from them, such as the quartile and the interquartile range;

    draw and read histograms with unequal intervals.

STUDENT'S ANSWER - EXAMINER'S COMMENTS

QUESTION

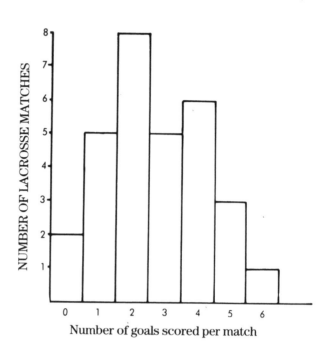

Number of goals scored per match

The above chart shows the goals scored per match in league lacrosse matches on a certain Saturday.

(i) Write down the number of matches in which 2 goals were scored.

Calculate
(ii) the number of matches played,
(iii) the number of goals scored altogether,
(iv) the mean number of goals scored per match.

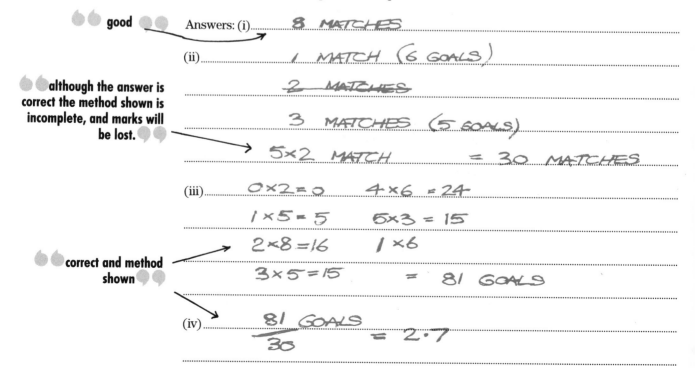

good Answers: (i)........ 8 MATCHES

(ii)........ 1 MATCH (6 GOALS)

although the answer is correct the method shown is incomplete, and marks will be lost.   2 MATCHES

3 MATCHES (5 GOALS)

5×2 MATCH   = 30 MATCHES

(iii)........ 0×2=0   4×6 =24

1×5 = 5   5×3 = 15

2×8 =16   1 ×6

correct and method shown   3×5 =15   = 81 GOALS

(iv)........ 81 GOALS / 30   = 2·7

I N D E X